SACRED FORESTS
Their Ecology and Diversity

SACRED FORESTS
Their Ecology and Diversity

by
H.N. Pandey
Professor & Ex-Head
Department of Botany
North-Eastern Hill University
Shillong

2010
REGENCY PUBLICATIONS
Delhi - 110 035

ISBN 81-89233-64-5
ISBN 978-81-89233-64-8

Published by : **REGENCY PUBLICATIONS**
(Division of Daya Publishing House)
4760-61/23, Ansari Road, Darya Ganj,
New Delhi - 110 002
Phone: 23245578, 23244987
Fax: (011) 23260116
e-mail : info@regency-books.com
website : www.dayabooks.com
website : www.regencybooks.com

Laser Typesetting : **Classic Computer Services**
Delhi - 110 035

Printed at : **Chawla Offset Printers**
Delhi - 110 052

PRINTED IN INDIA

Foreword

It gives me immense pleasure in writing this foreword. Local communities in many parts of the world have conserved sacred forests or groves based on spiritual and cultural values. This ancient multi-dimensional traditional institution has been studied among others, by social scientists, biologists and ecologists. The studies have generated considerable data and knowledge, in particular regarding number and distribution of sacred groves, associated belief systems, role it plays in the lives of local communities, its conservation status etc. However, there are not many works that have studied in depth ecological dimensions of sacred groves.

Prof. H.N Pandey, his colleagues and students at North-Eastern Hill University (NEHU), Shillong have studied several groves in the state of Meghalaya covering ecological dimensions such as floristic diversity, community structure, tree regeneration and nutrient dynamics.

The author in this book, besides giving an overview of sacred groves in India, provides comprehensive synthesis of studies carried out at NEHU. The book has significantly advanced our knowledge on ecological aspects of sacred groves. The findings of Prof. Pandey's studies that sacred groves harbour rich bio-diversity including rare, threatened and endangered species and are self sustainable ecosystems, further reinforces scientific reasons for conserving sacred groves.

I am confident that Prof. Pandey's work will inspire young researchers to carry out further studies on sacred groves covering socio-cultural and biological dimensions. The book will be of immense use to researchers, local communities, Forest Departments and NGOs.

Lastly, I would like to congratulate Prof. Pandey and his colleagues for making this outstanding contribution in the growing literature on sacred groves.

Kailash C. Malhotra

FNA

Preface

Indigenous and local communities all across the globe have conserved thousands of sacred sites and sacred forests for millennia for spiritual, cultural, economic and aesthetic purposes. More recently when the species are getting extinct at faster rate due to forest destruction all over the world, sacred groves are serving safe abode for a large number of rare, endemic and threatened plant and animal species and playing impotent role in providing essential environmental and economic benefits to the local communities. This has led to greater attention of social scientists, biologists, non-government and voluntary agencies in the study of different aspects of sacred grove institution all over the country. As a result large number of general articles, research papers and books on different aspects of sacred groves of India has been published during the past three decades or so. Majority of these deal with the listing of groves, documentation of traditional beliefs and cultural practices of different indigenous communities associated with sacred groves, their roles in social life, status and management of sacred groves, their importance in biodiversity conservation and forces threatening existence and suggestions and strategies for conservation of sacred gove etc.

The plant taxonomists, ecologists and anthropologists have made valuable contribution to our knowledge of the sacred groves found in different parts of the country. Large number of studies dealing with documentation, floristic composition, community structure, conservation etc., has been carried out in western and southern India where more than fifty percent of the total documented sacred groves of the country are found. The most intensive ecological studies of sacred grove ecosystem have been carried out in the northeastern states of India. These include analysis of species diversity, structure and dynamics of plant communities, population structure of tree species and their regeneration behaviour, litterfall and decomposition, fine root production, accumulation and decomposition, soil microbial biomass and mineralization and soil nutrient status and dynamics, and several other community-level processes that make sacred groves species-rich self sustainable ecosystem.

The plant ecology research group in the Department of Botany at North-Eastern Hill University, Shillong, has made significant contribution to the understanding of sacred grove ecosystem found in the state of Meghalaya. Eleven (Appendix Table 1) groves distributed between 1300 and 2000 m altitude in the state representing the subtropical broadleaved semi-evergreen forest were intensively studied by doctoral and post-doctoral researchers during the past two decades. Their published and unpublished research findings and those of other workers engaged in ecological study of sacred groves of the country form the basis of this book.

This book is very much a product of research of students and colleagues who have ben associated with me at the Department of Botany, North-Eastern Hill University, Shillong during the past 20 years. I should like to thank them for their cooperation and help in completion of the project. Among them my colleague Prof. R.S. Tripathi was a constant source of inspiration and encouragement. I take this opportunity to express my grateful thanks to him.

I am grateful to Prof. B.K.Tiwari, Coordinator, Regional Centre, National Afforestation and Ecodevelopment Board, Ministry of Environment and Forest, Government of India, NEHU, Shillong for his valuable help and support in the execution of the project. I express my deep sense of appreciation and gratefulness to Prof. S.K.Barik, Department of Botany, North-Eastern Hill University for his cooperation, valuable suggestions and help in completing the project.

My students Dr. Krishna Upadhaya and Dr. (Ms) Jenpuiru Kamei, North-Eastern Hill University, immensely helped me in compilation of data and development of chapters on floristic diversity, community structure and nutrient dynamics. I express my grateful thanks to both of them. In addition, I am grateful to all authors whose research findings are cited in different chapters of the book, and would be pleased to be notified of any omissions.

I thankfully acknowledge the help recieved from the research students of Ecology group at the Department of Botany, NEHU, especially Dr. Dibyendu Adhikari in copy-editing of different chapters of the book.

I express my grateful thanks to Prof. K.C. Malhotra for his valuable intellectual input during the preparetory phase of the book and for writing its foreword. I am grateful to Prof. J.S. Singh, Secretary, Banaras Hindu University, for his timely help during submission of the project proposal to the Department of Science and Technology.

I extend my heartful thanks to my wife Smt. Sudha Pandey for her encouragement and support during the project duration.

The project was catalyzed and supported by the Department of Science and Technology, Government of India, New Delhi under its "Utilization of Scientific Expertise of Retired Scientists Scheme" as per the sanction order No.HR/UR/18/2006, dated 29.01.07. I thankfully acknowledge the financial assistance received from the Ministry for the project.

H.N. Pandey

Contents

List of Illustrations

List of Tables

List of Text Tables

List of Appendix Tables

Chapter 1

Sacred Forests

1.1 The Concept

Conservation of plants and other natural reserves has been an integral part of cultural ethos of indigenous communities in many parts of the world who considered themselves as being linked in a web of spiritual relationship with their biophysical environment (Huges and Chandran 1998). Conservation was the result of nature worship of the indigenous people all over the world. They considered specific plants or animals or even rivers and mountains as their ancestors and protected them. The evidence for this is seen even today in the rituals of many hunter gatherer groups. There are symbolic inclusion of wild animals, hunter gods and goddesses and hunting weapons in the rituals associated with the worship in the sacred groves. Though hunting was usually forbidden in the holy precincts, it was sometime permitted for an offering to a resident god or goddess. In India many societies or cultures have traditionally developed strategies for conserving and managing nature and natural resources. These strategies were highly congruent to the traditional lifestyle of the respective societies, and were determined by human attitudes towards utilization of the resources.

In India nature worship dates back to the Vedic period (5000 BC), and is based on the premise that all creations of nature have to be protected. Several plant species such as Peepal (*Ficus religiosa*),

Tulsi (*Ocimum sanctum*), Rudraksha (*Eleocarpus* sp.), Bel (*Aegle marmelos*), Ashok (*Saraca asoka*) and Kadam (*Anthrocephalus chinesis*) have been considered sacred. Besides sacred species, species required for rituals, and patches of forests and water bodies were also conserved in the name of local deities. An important tradition of nature worship is to protect patches of forest dedicated to deities or ancestral spirits. These forest patches have been designated as '*Sacred groves*'. They manifest the spiritual and ecological ethos of local indigenous communities. Recently efforts have been made by the scientists to define the sacred groves. One such effort is that of Hughes and Chandran (1998) who have defined sacred groves as "segments of landscape, containing vegetation, life forms and geographical features, delimited and protected by human societies under the belief that to keep them in a relatively undisturbed state is expression of an important relationship of humans with the divine or with nature".

It is believed that the existence of sacred forest dates back to several thousands of years when human society was in the primitive stage of development. Gadgil and Vartak (1975) have traced the historical link of the sacred groves to the pre-agricultural, hunting and gathering stage of societies, before human being had settled down to raise livestock's or till land. Hence, the concept of virgin forest is believed to be of pre-Vedic period, *i.e.*, about 3000 to 5000 years BC.

Hughes (1984) traced the origin of the concept of sacred groves in European and Greek Roman context and argued that the sacred groves or sanctuaries were the first temples of God. In Greece groves and forests were enclosed usually by stone walls. This enclosure was called *Temenos* in Greek, meaning a cut-off place or a demarcated place.

It is also believed that shifting cultivation could be one of the reasons for creation of sacred groves. In this most primitive form of agriculture in forested regions patches of forests might have been spared from slashing and burning, and have been left for the local deities in the form of sacred groves (Gadgil and Vartak 1976, Hazra 1975, 1980). Since these forest patches were often protected watersheds with a perennial source of water, people believed that the deity residing in these forests yields water. This is considered as the most popular belief of the origin of sacred groves.

Sacred groves might have also originated as a result of its utilitarian nature (Gadgil and Vartak 1976), a social institution (Kalam 1996) or as a part of the taboos that evolved historically over several generations to provide a site for culturally crucial social interactions (Gerden and Mtallo 1990). The ancestral practices of animism with the central focus on the worship of forest patches regarded them as the sacred abode of various gods/deities. Thus nature worship has been a key force in determining human attitudes towards conservation and sustainable utilization of natural resources.

Sacred groves throughout the world are associated with a range of traditional and cultural values related to forests, rituals and taboos. However, conservation practices and control over extractive activities in sacred groves vary in different communities and regions.

1.2 Distribution of Sacred Forests in the World

The sacred groves, sacred areas, sites and geographies are found almost all over the world. The sacred groves have been reported from different continents of the world such as Africa, Asia, Europe, Austro-Pacific region and Americas (Hughes and Chandran 1998, Malhotra *et al.*, 2007). In Africa it is found in Sierra Leone, Ghana, Ivory Coast, Nigeria, Old Calabar, Zimbabwe, South Africa, Egypt and Kenya. In Asia the sacred groves have been reported from India, Korea, Japan, China, Thailand and Indonesia. Countries like Germany, Britain, Italy and Finland in Europe had thousands of sacred groves in ancient times. Most of them have disappeared now. The sacred groves are widely known in the Austro-Pacific Region, New Zealand and Polynesia. In America native people believed that although the whole earth is sacred, but in certain places the spirit power manifests itself more clearly and readily. Many of these places were groves of trees. Americans of European of descent also have sometimes hallowed groves of trees. Some of them have been rededicated to early settlers, or to the war deads, or to noted leaders.

1.3 Distribution of Sacred Forests in India

In India, the sacred groves are found all over the country, especially in the regions inhabited by indigenous communities, particularly along the Western Ghats in the states of Maharashtra, Kerala, Karnataka and Tamil Nadu. In north-east India most of the sacred groves has been reported from Arunachal Pradesh,

Meghalaya and Manipur (Figure 1.1). Although, there has been no comprehensive survey of the sacred groves in the entire country, approximately 13,720 sacred groves have been documented so far (Appendix Table 2). Experts estimate the actual number could be much higher in the range of 100,000–150,000.

The earliest documented work on sacred groves in India is that of Brandis (1897a, b), the first Inspector General of Forests. Sacred groves reported from different parts of India are locally known by different names for example, sacred groves are called as *'Sarna'* or *'Dev'* in Madhya Pradesh (Bhakat 1990), *'Devrai'* or *'Devrahati'* in Maharashtra, *'Sarnas'* in Bihar, *'Orans'* in Rajasthan, *'Devarabana'* or *Devarakadu'* or *'Rulidevarakadu'* or *'Nagabana'* etc. in Karnataka, *'Kovilkadu'* in Tamil Nadu, *'Kavu'* in Kerala, *'Dev van'* in Himachal Pradesh, *'Ki Law Lyngdoh'* or *'Ki Law kyntang'* etc. in Meghalaya, *'Sarana'* or *'Jaherthan'* in Jharkhand and *'Lai Umang'* in Manipur.

Gadgil and Vartak (1975, 1976) prepared an inventory of the sacred groves or *'Devrais'* of the state of Maharashtra. Detailed information about the location, area and associated deity, folklores and traditional beliefs of 233 groves from different districts of Maharashtra were collected by Gadgil and Vartak (1981). According to Burman (1992), the number of sacred groves in Maharashtra and in the Western Ghats may be more than the number recorded by Gadgil and Vartak (1976). Balasubramanyam and Induchoodan (1996) reported 761 sacred groves in Kerala. Out of these 399 were of less than 0.02 ha in extent (Induchoodan, 1996). Perhaps highest density of sacred groves in the world is found in Karnataka (Kushalappa *et al.*, 2001). Kadamba (1998) enumerated 80 sacred groves from south-eastern coast encompassing the Marakkanam-Pondicherry-Cuddalore regions. In Andhra Pradesh 425 sacred groves were reported from Anantapur, Kurnool, Prahasam, Nellore and Cuddapah districts (Anonymous 1996). Basu (2000) reported a sacred grove from the Purulia district of West Bengal, which has an area of over 72,681m^2. Panda *et al.* (2003) reported 10 important sacred groves of Santhals from the Bankura district of West Bengal. Tiwari *et al.* (1999) inventoried 79 sacred groves of Meghalaya. Devi (2000) reported 365 sacred groves from Manipur. Dedicating a patch of forest to deities is a common practice with the Meitei community of Manipur who follow ancestral practices of animism with the central focus on the worship of forest patches which are regarded as sacred abode of *Umanglais* (sacred deities or sylvan-deities).

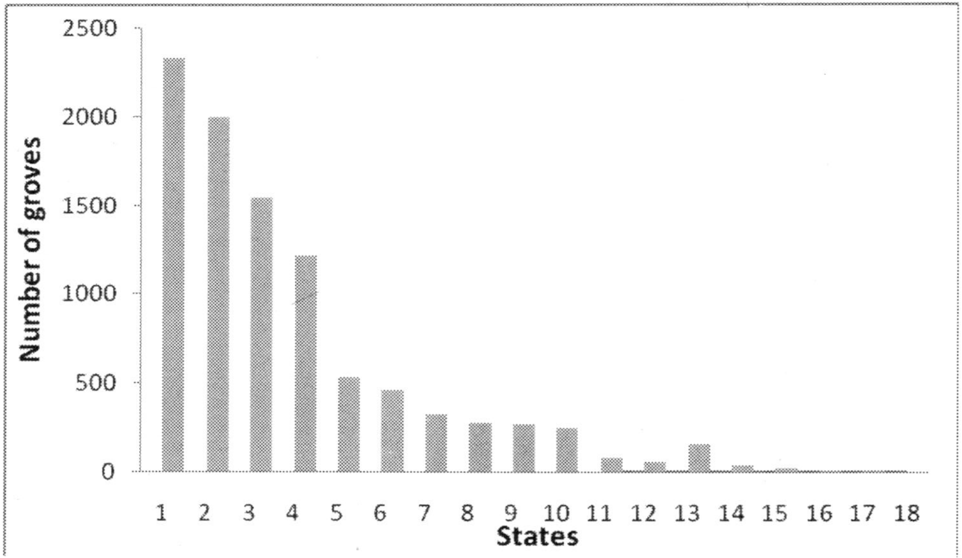

Figure 1.1: Distribution of Sacred Forests in Different States of India

(1. Maharashtra, 2. Kerala, 3. Andhra Pradesh, 4. Karnataka, 5. Manipur, 6.Tamil Nadu, 7. Orissa, 8. Madhya Pradesh, 9. West Bengal, 10. Haryana, 11. Meghalaya, 12. Sikkim, 13. Arunachal Pradesh, 14. Assam, 15. Gujarat, 16. Himachal Pradesh, 17. Rajasthan, 18. Uttar Pradesh)

The size of the individual sacred grove varies from a clump of a few trees to several hectares. Some sacred groves like those reported from Purulia district in West Bengal and Ri-Bhoi and East Khasi Hills districts in Meghalaya are spread over several hundred hectares (Basu 2000, Tiwari *et al.*, 1999). In Manipur majority of the groves are between 0.09 ha and 1.5 ha area. The total area of all inventoried sacred groves in Manipur is 756.42 ha.

The estimated number of sacred groves reported from different parts of the country is 13, 720 (Malhotra *et al.*, 2001a, Appendix Table 2), but actual number is likely to be much larger varying between 100, 000 and 150, 000. For instance, recently in Meghalaya about one hundred sacred groves have been reported by Barik *et al.* (2006). But this number is far less than the actual number of groves that might be present in Meghalaya, since every village in the Khasi-Jaintia hills is thought to have one or more sacred groves.

1.4 Classification of Sacred Forests

Attempts have been made to classify the sacred groves found in India. In one such effort three categories have been recognized based on their religious, cultural and other values for which they are used.

- ✰ Traditional Sacred Groves–Places where the village deity reside and are represented by an elementary symbol.
- ✰ Temple Groves–Groves created around a temple and conserved.
- ✰ Groves around the burial or cremation grounds.

1.5 Ecological Importance of Sacred Forests

The concept of sacred groves and sacred landscapes has been the focus of academic discussion and research among the biologists and anthropologists during the last four decades. Their works have contributed immensely in increasing the awareness about the significance of sacred groves in biodiversity conservation. Since sacred groves are remnants of the original forests, it is expected that at least a portion of the original biodiversity has survived in these forests. Some plant species growing in the sacred groves are used by traditional healers and priests who have a strong interest in the preservation of such sites and ecosystems. Small farmers, fish folk, herders and hunter gatherers, depend on the diversity of species, genetic varieties and ecosystem services for their livelihood and cultural lives. These aspects of sacred groves have been explicitly discussed by Norse and McManus (1980).

Apart from the direct benefits of biodiversity conservation, indigenous societies have always sensed the indirect benefits which flow from the sacred groves in a variety ways through ecosystem functions. To understand the relationship between biodiversity and ecosystem function is a major challenge to the scientific community. An array of interactions between physical and biological components of the sacred grove ecosystem help in moderation of climatic regimes, regulation of local hydrology, maintenance of nutrient cycles and soil fertility and provide aesthetic scenic beauty to the landscape etc., which are vital to humans. In a short time span biodiversity affects several community level processes within the ecosystem and in the long-term contributes to the stability and maintenance of ecosystem processes in face of perturbation (Loreau 2000). Large numbers of species of diverse life-forms with their differing ecological amplitudes and requirements and functional attributes in the community contribute to the stability and perpetuation of sacred grove ecosystem. Some species though present in small numbers often play very important roles in functioning of the ecosystem by their unique abilities such as nitrogen fixation etc. Such species are variously known as keystone species or functional groups. The roles of such species in regulating the ecosystem functions are being investigated by several workers (Schulze and Mooney 1993, Tilman *et al.*, 1997, Hooper and Vitousek 1997).

Few decades ago the sacred groves exemplified a self sustainable ecosystem due to minimum human intervention. The biological communities in the groves were in dynamic equilibrium with the prevailing climatic and edaphic conditions and represented the terminal stage of development under that condition. The delicate ecological balance that evolved as a result of innumerable intricate interactions both within the community as well as between the community and its physical environment through ages has been disturbed due to biotic stress of varied kinds and magnitudes primarily caused by human activities. As a result, a large number of groves found in different parts of the country are no longer in a state of dynamic ecological equilibrium. The plant species which form the major structural component of the sacred grove ecosystem have been changed, the soils which support the vegetation have been degraded and the forest microclimate that is crucial for regeneration of tree species have been altered. These changes have upset the ecological balance and have disrupted the structure and functioning of the majority of sacred grove ecosystems which no longer exist in self-sustainable state (Figure 1.2).

1.6 Socio-cultural Importance

In addition to the ecological importance, the sacred groves have important socio-cultural functions to the indigenous communities. The relationship between cultural and biological diversity in India has evolved as a result of close relationship between the two. The biodiversity has been conserved and survived against the forces of destruction due to strong bondage between socio-cultural values attached to the sacred groves which happen to be the store house of biodiversity. Some of the socio-cultural values of the sacred groves are summarized in Table 1.1.

Table 1.1: Socio-cultural Values of the Sacred Groves

Religious: Propitiations of deity/ancestral spirits/totems.
Cultural value: Provides cultural space to the community as a common property resource. Assertion of group identity/group solidarity/new alliances etc.
Health: Well being of community/family/individual.
Economic value: Good rainfall/good agricultural production/well being of crops and animals/source of raw materials, medicine, food, fruits etc.
Psychological: Moral support and guidance to the local people.
Educational: Ideal sites for teaching and research.

The age-old cultural ethos of tribal communities which were responsible for preservation of the integrity of the sacred grove through ages is degrading at a fast rate. The traditional belief systems that were fundamental to the concept of sacred grove conservation are being considered as mere superstitions and are gradually disappearing with the advent of modernization, urbanization and people's changing aspiration. Modernization is posing a serious threat to the local traditions which are being challenged by urban cultures and the sacred groves are losing their cultural importance for the younger generations of the people. There are instances where change in religious beliefs of people is responsible for the degradation of sacred groves.

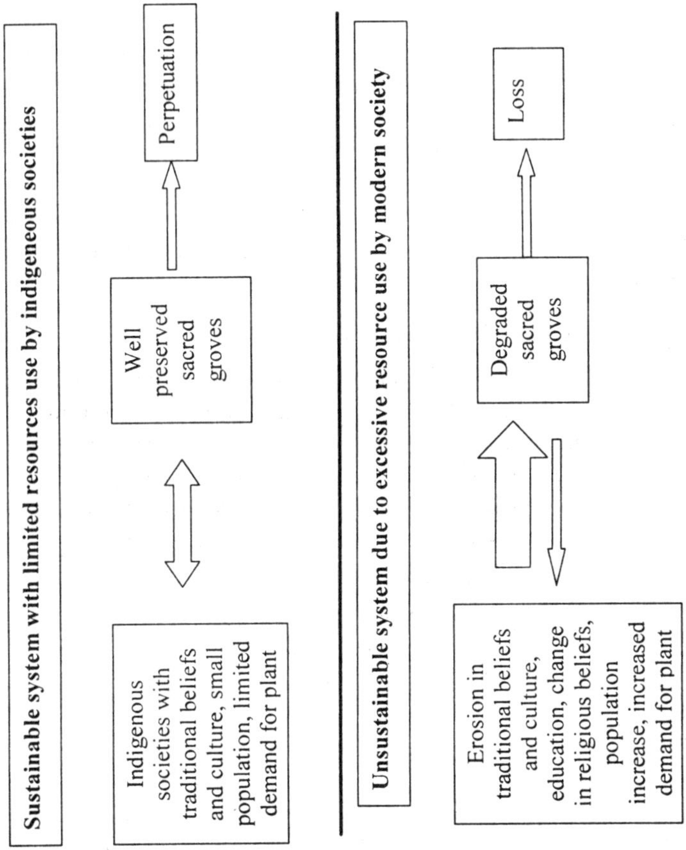

Figure 1.2: Schematic Diagram Showing Linkages Between Sacred Grove, Tribal Community and Factors Responsible for Sacred Grove Degradation

The shrinking primary forests are also forcing the local communities to encroach the sacred groves. Human activities that were previously taboo, such as dead wood collection, lopping of trees for timber and fuel wood, collection of wild edible fruits and tender branches and green leaves for fodder and cattle grazing have increased. Many sacred groves are fragmented by construction of footpaths, roads, extension of power lines etc. The disturbed and fragmented sacred groves are being invaded and colonized by exotic weeds like *Eupatorium adenophorum, Lantana camera, Artemisia nilagirica* etc., threatening the indigenous species present in the groves. The loosening of management of sacred groves by traditional institutions is also a major factor contributing to the sacred grove degradation. There are instances where decisions were taken to benefit certain sections of the village society against the traditions of the sacred groves.

1.7 Degradation of Sacred Forests

The sacred groves distributed in different parts of the country are facing various kinds of anthropogenic stresses. The intensity and frequency of these biotic stresses vary from one region to the other and even from one grove to the other in the same region. The commonly recognized factors which are responsible for destruction and degradation of the sacred groves are summarized below.

- ☆ Loss of faith in the traditional belief systems, which were fundamental to the concept of sacred groves. These systems and the rituals attached to the belief system are now considered mere superstition.
- ☆ Many groves are suffering due to transformation of the primitive forms of nature worship into formal temple worship. In many parts of the country change in the religious beliefs from traditional/Hindu way of worship to Christianity has led to destruction of the sacred groves.
- ☆ Sacred groves in many parts of the country have been destroyed due to rapid urbanization and developmental interventions such as roads, railways tracks, dams and commercial forestry. Encroachment has led to the shrinkage of some of the largest groves in the country.
- ☆ Collection of biomass in the form of green manure, medicinal herbs, tree branches and dead wood as well as

extraction of timber with religious justification is also contributing to degradation of the sacred groves.

- ☆ Invasion by exotic weeds such as *Eupatorium odoratum, Lantana camara and Prosopis julifera* etc. is a serious threat to some groves.
- ☆ Grazing pressure due to increasing livestock population and fuel wood collection is also responsible for sacred grove destruction in some areas.
- ☆ Land reforms in some states, for example in Kerala, have led to reduction in large land holdings of big land owners. Therefore they could not afford to have sacred groves in the remaining portion of their land, which in due course of time was brought under other uses.

Chapter 2

Meghalaya: An Overview

The state of Meghalaya lies between latitude 25° 02' and 26° 07' N, and longitude 89° 49' and 92° 50' E with an area of 22, 429 square km. The topography of the state is variable with elevation ranging from 60 m to 2040 m asl. Being a part of the northeastern Indian bio-geographic zone, it constitutes the meeting place of paleo-arctic, Indo-Malayan and Indo-Chinese bio-geographic realms. The state has high concentration of plant species and harbors about 3128 species including 1237 endemic flowering plants in just 0.7 per cent of the country's land area (Khan *et al.*, 1997).

2.1 Geology

Meghalaya represents the remnant of the ancient plateau of Pre-Cambrian Indian peninsula and forms a prominent geomorphic unit stretching across the Garo, Khasi and Jaintia hills in east-west direction. The central and northern part of Meghalaya plateau is made essentially of highly metamorphosed crystalline rocks of Pre-Cambrian origin, which are often referred to as "Archaean Gneissic Complex" (Anonymous 1974). It is made up of gneisses (biotic gneiss, biotite granulites, quartz-sillimanite gneiss, cordierite, garnet and chondrodite) and Schistose (mica schist, quartz-sillimanite schist and metabolite) members of varying composition.

The Meghalaya-Karbi plateau are linked with Chotanagpur plateau with an under grouped extension below the Raja Rajshahi-

Rongpur gap (Rajmahal-Garo Gap or Malda Gap). The massif is a geomorphologic arch bounded on all sides by faults (Majumdar 1986). The southern and northern boundaries are marked by the Dawki fault and Brahmaputra lineament, respectively. The west massif is bordered by Rajmahal-Garo lineament and eastern boundary is a NE-SW lineament separating the massif from the sediments of Bengal-Assam shelf. The southern parts of the Shillong plateau have experienced tremendous tectonic impacts giving rise to the steep scarp which today stands aloft facing the Sylhet plain of Bangladesh (Singh 1996).

2.2 Physiography

Based on the physiography, the state may be divided into (i) western region, (ii) central and eastern region, (iii) northern undulating hills, and (iv) southern precipitous zone. The western region includes Garo hills and western parts of west Khasi hills. The central and eastern region includes Khasi hills, Ri-Bhoi and Jaintia hills. The northern region is characterized by undulating hills towards the north with altitude ranging between 300 m and 1600 m asl, while the southern part is a high precipitous zone. The central upland zone constitutes the highest elevation zone of the Shillong plateau.

2.3 Climate

The climate of the state is directly influenced by the southwest monsoon originating from the Arabian Sea and the Bay of Bengal. The climatic variables like temperature, rainfall and humidity vary widely from place to place in the state due to wide variation in topography. Based on the atmospheric conditions, the year may be divided into summer, rainy, autumn and winter seasons. The summer (March end to mid May) is characterized by warm weather with occasional thunderstorms and high wind velocity. The rainy season commences with the onset of southwest monsoon in mid May and continues up to September. Three-fourth of the total annual rainfall is received during this season. The rainy season is followed by a brief autumn during October and November. Rainfall and temperature sharply decline during this period. The winter season, extends from December to February. Morning fog and frost, and dry weather are the characteristic features of this season.

The annual rainfall varies widely in the state ranging from about 1,600 mm (Umiam, Ri-Bhoi) to over 11,463 mm (Mawsynram, East Khasi hills). The Ri-Bhoi district receives the lowest rainfall, while the northern part of Jaintia hills, Khasi hills and whole Garo hills receive between 1,600 and 4,000 mm rainfall annually where altitude ranges between 600 m to 1400 m asl. In southern Khasi hills at higher altitude, annual rainfall varies widely from 4,000 mm to 11,000 mm at different places.

The western part of the state is warm with mean temperature ranging between 12°C and 33°C. The central upland is relatively cooler where the mean minimum temperature drops to 5 °C and mean maximum temperature does not exceed 30 °C. The climatic data for the five different meteorological station *viz.*, Tura (Garo hills), ICAR complex (Ri-Bhoi district), Upper Shillong and Cherrapunji (East Khasi hills), and Jowai (Jaintia hills) for the period 1997- 2001 are shown in Table 2.1.

Table 2.1: Average Annual Rainfall (mm), Maximum and Minimum Temperatures (°C) and Relative Humidity (per cent) at Different Stations in Meghalaya

Parameters	*Stations*				
	Tura	*Umroi*	*Cherrapunji*	*Jowai*	*U. Shillong*
Altitude (m asl)	350	1010	1350	1450	1900
Rainfall (mm yr^{-1})	4321	2352	11995	5701	3268
Maximum Temperature (°C)	31.7	22.9	20.9	23.4	17.7
Minimum Temperature (°C)	16.5	14.9	13.5	13.8	10.1
Relative Humidity (per cent)	66.4	78.1	78.5	66.9	76.8

2.4 Soil

The soils of Meghalaya have been broadly grouped into four categories *viz.*, (i) red loamy soil, (ii) lateritic soil, (iii) red and yellow soil and (iv) alluvial soil. The red loam soil is found in the central part of the Garo Hills and on the uplands of central and eastern Meghalaya. The lateritic soil occurs in the form of a broad belt that extends from west to east in the northern part of the state. Red and yellow soil extends from west to east in the southern part. Alluvial soil occurs all along the northern, western and southern fringes of the state. The soils are acidic (pH 3–6.5) in nature. Acidity is attributed to the leaching of cations like calcium, magnesium and potassium from the soil due to high rainfall and undulating topography. Total Kjeldahl nitrogen content is low (*ca.* 0.15 per cent) at degraded sites and fairly high (*ca.* 0.94 per cent) in the undisturbed forests. Soils are deficient in available phosphorus ($< 15\ \mu g\ g^{-1}$) and medium in available potassium (*ca.* 13 $mg\ g^{-1}$), and rich in organic carbon (*ca.* 1.62 per cent). Soils are not suitable for intensive cultivation due to their poor base (*ca.* 35 per cent) saturation.

2.5 Vegetation

One of the most important forest classifications of greater India was given by Champion (1936) and later on revised by Champion and Seth (1968). The revised classification recognizes 16 major forest types mainly on the basis of climatic factors and subdivided in to 221 minor types on the basis of structure, physiognomy and floristic composition. Puri *et al.* (1983) while broadly agreeing to their classification, have argued that biotic factors such as shifting cultivation, fire and grazing have played important role in determining development and influencing floristic composition and stability of the forest community in India, which ranges from evergreen tropical rain forests in the Andamans and Nicobar Islands, the Western Ghats, and the northeastern states to dry alpine scrub in the Himalayas. Between the two extremes the country has semi-evergreen rain forests, deciduous forests and thorn forests, subtropical forests in the lower montane zones and temperate montane forests (Lal 1989). The semi-evergreen forests are more extensive than the evergreen formations partly because the later tend to degrade in to semi-evergreen type with human interference.

The tropical vegetation of northeast India, which includes the states of Assam, Nagaland, Manipur, Mizoram, Tripura and Meghalaya and plains of Arunachal Pradesh, typically occurs up to 900 m elevation. It includes evergreen and semi-evergreen rain forests, moist deciduous forests, riparian forests, swamps and grasslands. The evergreen forests are found in Assam valley, foot hills of eastern Himalayas and lower parts of Naga Hills, Meghalaya, Mizoram and Manipur where annual rain fall exceeds 2000 mm.

The vegetation of Meghalaya has been classified by Kanjilal *et al.* (1934-40), Champion and Seth (1968), Balakrishnan (1981-83), Haridasan and Rao (1985-87), Rao and Hajara (1986) and Chauhan and Singh (1992). They have divided the vegetation into tropical evergreen forest, tropical semi-evergreen forest, subtropical broad-leaved hill forest, tropical moist deciduous forest, temperate and subtropical pine forest and savanna.

Table 2.2: Forest Types of Meghalaya

Champion and Seth (1968)	Tropical evergreen forest
	Tropical semi-evergreen forest
	Subtropical broad-leaved hill forest
	Tropical moist deciduous forest
	Temperate forest
	Subtropical pine forest
	Savanna
Balakrishnan (1981-83)	Tropical evergreen forest
	Tropical semi-evergreen forest
	Subtropical evergreen forest
	Subtropical semi-evergreen forest
Haridasan and Rao	Tropical evergreen forest
(1985-87)	Tropical semi-evergreen forest
	Tropical moist and dry deciduous forest
	Temperate forest
	Subtropical pine forest
Chauhan and Singh (1992)	Tropical forest
	Subtropical forest
	Temperate forest
IIRS (2000)	Tropical moist deciduous forest
	Subtropical evergreen forest
	Subtropical semi-evergreen forest
	Subtropical pine forest

National Remote Sensing Agency (NRSA) has used satellite remote sensing technique to map forest cover of India, including Meghalaya (Anonymous, 1984). The Forest Survey of India (FSI) has been mapping forest cover of the state since 1987. The vegetation cover of the state in 1999 was 15,633 square km, or 69.7 per cent of the total geographical area of the state (Anonymous 1999). Dense forest occupied 5,925 square km (26.4 per cent) and open forest occupied 9,708 square km (43.3 per cent) of the vegetation cover. The biosphere reserve, reserve forests, protected forests, national parks and wildlife sanctuaries together covered about 300 square km area. Based on the IRS ID data of 2000, six major forest types have been recognized in the state (Map 2.1). They together cover 42.2 per cent of the geographical area of the state. Out of these, the subtropical evergreen forest, subtropical semi-evergreen forest, tropical moist deciduous forest, and subtropical pine forest cover 11.6 per cent, 21.5 per cent, 1.3 per cent and 7.8 per cent of the total geographical area, respectively.

The tropical forest occurring below 1000 m is either evergreen or semi-evergreen type depending on the dominance of evergreen and deciduous trees in the canopy. The subtropical forests found above 1000 m are either broad-leaved or needle-leaved. Small pockets of subtropical evergreen broad-leaved forest are found where rainfall is relatively high and soil moisture condition remains favorable for most part of the year. Those areas which receive relatively low annual rainfall support semi-evergreen forest. Pine forests have developed as a stable secondary community on the disturbed evergreen and semi-evergreen subtropical broad-leaved forest sites, which are seasonally dry and have nutrient-poor soil.

(*i*) Tropical Moist Deciduous Forest

This forest type is represented by sal-bearing forest at low elevation in Rongrengiri, Sangsuk, Darugiri and Baghmara areas of Garo hills where annual rainfall is less than 150 cm. The forest covers a small area of about 124.8 sq. km in the state. Along with *Shorea robusta*, other tree species like *Bauhinia variegata*, *Callicarpa arborea*, *Cordia grandis*, *Calliandra* sp., *Dillenia scabrella*, *Dysoxylum binectariferum*, *Embelia floribunda*, *Lagerstroemia parviflora*, *Mallotus philippensis*, *Picrasma javanica*, *Sterculia villosa*, *Styrax serrulatum*, *Schima wallichii*, *Tectona grandis* and *Terminalia myriocarpa* are found in the forest. The understorey is composed of *Clerodendrum viscosum*, *Eupatorium adenophorum*, *Psychotria monticola*, *Melastoma malabathricum*, *Pongamia* sp., *Sabia purpurea*, *Ardisia neriifolia*, *Digitaria* sp., *Desmodium* sp., *Gleichenia* sp., and *Vandelia* sp.

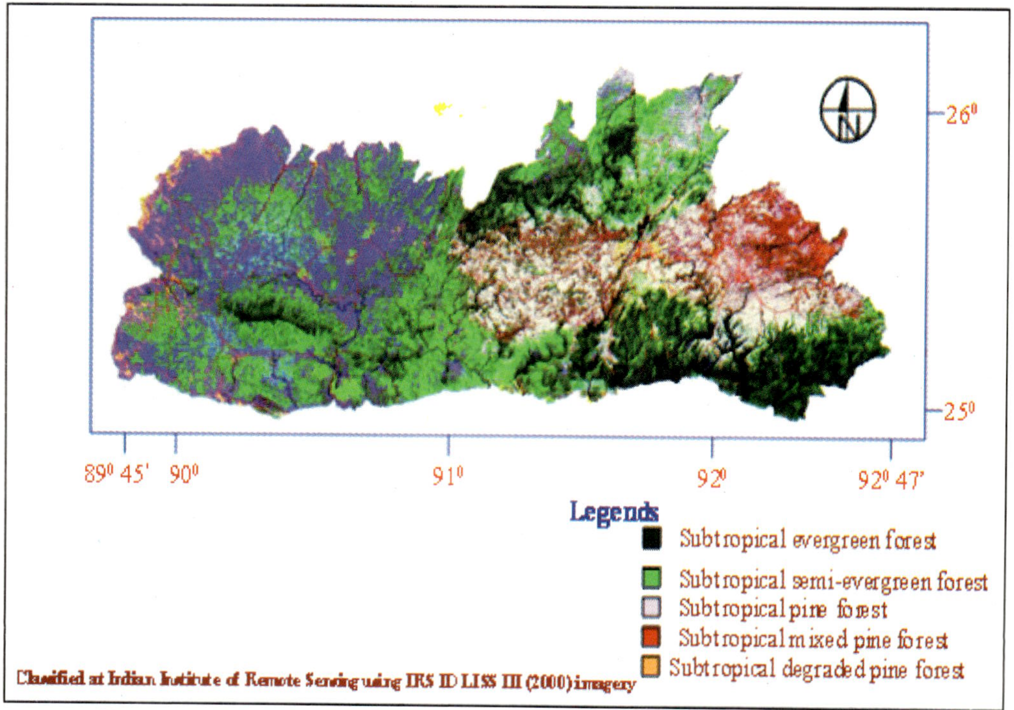

Map 2.1: Vegetation Map of the State of Meghalaya Showing Forest Types

(*ii*) Subtropical Evergreen Forest

It generally occurs above 1200 m asl where average annual rainfall ranges between 300 and 500 cm and temperature shows a noticeable difference between summer and winter season. The ground frost is common in December-January. The forest covers about 2547 sq. km in Jaintia hills (some parts of Narpuh and Saipung reserve forest), Khasi hills (Mawphlang and Cherrapunji areas), Ri-Bhoi (Langbi, Kariong, Umter, Mopon and Nogpydem areas), and Garo hills (upper reaches of Tura and Siju reserve forest) districts. It is found in patches of varying sizes ranging from < 1 sq. km to 110 sq. km mainly on inaccessible hill slopes and valleys along the banks of rivers and streams. Large continuous patches are rare.

The trees are generally short statured not exceeding >25 m height. Buttressed trunks and lianas are rare. Stratification is indistinct in the valleys, but it is clear at hilltops. The shrubby and herbaceous layers are clearly seen in the forest. Epiphytes, mosses and liverworts are abundant. The forest floor is spongy due to presence of thick litter and duff layers and a dense network of fine roots. Tall scattered trees of *Betula alnoides, Castanopsis* sp., *Exbuklandia populnea, Engelhardtia spicata, Ficus elastica, Lithocarpus elegans, Mangleitia insignis,* and *Prunus nepalensis,* constitute the canopy layer of the forest. Occasionally, *Schima wallichii* is also seen. *Viburnum foetidum, V. simonsii, Quercus glauca, Helicia nilagirica, Michelia punduana, Vernonia vulkamerifolia, Daphne involucrata, Symplocos racemosa* and *Ligustrum robustum* are common in the sub-canopy layer. Tree ferns are commonly found in the forest.

The common shrubby species found in the forest include *Ganinothalamus sesquiopedalis, Sarcococca saligna, Sarcandra glabra, Baliospermum micranthum, Neillia thorsiflora, Ixora subsessilis, Clerodendrum* sp., *Eurya japonica, Psychotria* sp., *Ardisia* sp., *Camellia caudata,* and *Saurauria* sp., and members of Acanthaceae and Araliaceae families. The forest floor is covered with mosses, selaginellas and angiospermous species such as *Begonia palmata, Senecio griffithii, Sonerilla* sp., *Impatiens* sp., *Didymocarpus punduana, Elatostema rupestre, Begonia rubro-venia, Disporum* sp., and members of Zingiberaceae, Araceae and Commelianaceae. *Clematis* sp., *Smilax* sp., *Dioscorea* sp., *Melodinus* sp. and species of Menispermaceae and Cucurbitaceae are the common climbers in the forest.

(*iii*) Subtropical Semi-Evergreen Forest

The altitudinal limits of distribution and climatic conditions prevailing in the subtropical semi-evergreen forest area are similar to those of evergreen forest. A transitional zone between tropical and subtropical forests is distinguishable at certain places between 1000-1400 m. These are found on northeastern slopes of Jaintia hills (major areas of Narpuh and Saipung reserve forest), Ri-Bhoi (Mopon, Mayang, Umsaw and Nogkhyllum wildlife reserve forest), Khasi hills (Mawsynram, Ryungud and Mosing, and southern parts of the district) and Garo hills (Damalgiri, Rangira, Sembu and Nokatgiri areas) and covers about 4788 sq. km area of the state. The forest is highly fragmented into small patches not exceeding 10 sq. km area due to timber extraction and shifting cultivation. There are few patches of larger size (>100 sq. km).

These forests are more species rich than the evergreen forest. Presence of few deciduous elements in the canopy which shed their leaves during the dry months between February and April, give semi-evergreen appearance to the forest. The number of deciduous trees, however, varies from site to site. Prickly and thorny species are commonly found in the forest.

The common canopy species are *Castanopsis indica, Diospyros undulata, Engelhardtia spicata, Elaeocarpus floribundus, Ficus altissima Sapindus rarak, Paramichelia baillionii, Meliosma wallichii,* and *Vitex glabrata.* The sub-canopy in the forest is composed of small trees of *Casearia vareca, Vitex vestita, Quercus semicarpifolia, Micromelum integrimum, Photinia arguta, Symplocos cochinchinensis* and *Xylosma controversum.* Lianas are less frequent and they are represented by *Mucuna macrocarpa, Tetrastigma obovatum* and *Celastrus campionii.* The shrub layer has fewer species such as *Crotalaria assamica, Boehmaria platyphylla, Capparis acutifolia, Lyonia ovalifolia, Randia griffithii, Mussaenda glabra, Desmodium* sp., *Maesa tetrandra* and *Clerodendrum* sp. The herbaceous layer is sparse. The common flowering plants of this layer are *Pilea umbrosa, Galinsoga parviflora, Anisadenia khasiana, Curcuma* sp., *Polygalla* sp., *Acanthus leucostachys, Pouzolzia hirta, Hedychium* sp., and many Asteraceae members. Mosses and ferns are rare.

(*iv*) Subtropical Pine Forest

The forest is confined to the central upland of Shillong plateau between 1000-2000 m asl. The climatic conditions are similar to those of evergreen and semi-evergreen forests. It occurs either as pure or mixed stands in Khasi and Jaintia hills above 800 m on nutrient-poor soils. The forest is exposed to annual winter fire when ground vegetation is almost completely dry. Besides annual fire, other biotic disturbances such as fuel wood collection, timber extraction and grazing are common in the forest. As a result of these activities, the forest has been fragmented into small patches. The area covered by this forest is about 1694 sq. km.

These forests occur on well drained, porous acidic (pH 4.65 to 5.12) soils with partially exposed to fully exposed rocks. Soil development is minimal where recurrent fire has consumed litter and under-story vegetation. There is usually very little organic matter left on the surface, which sometimes is a bare rock. The rocks are overlain with a thin layer of poor soil, which serves as the rooting medium for pine. Soils of the pine forests have relatively low organic carbon (1.2 to 3.2 per cent), organic matter (1.81 to 4.1 per cent), total Kjeldahl nitrogen (0.11 to 0.17 per cent) and available phosphorus (9.6 to 11.6 $\mu g\ g^{-1}$) than the soils of the semi-evergreen and evergreen forest stands.

The average height of pine trees ranges between 20 and 35 m, however, on degraded sites the height may be less. Few scattered trees of broad-leaved species are often associated with pine. A few small trees or large shrubs are found scattered in the forest forming the sub-canopy layer. Annual fire prevents establishment of shrubs and other woody elements. However, weeds and perennial grasses form dense undergrowth during monsoon. These forests are very poor in tree species content. At places it forms a mixed stand with *Schima wallichii, Prunus undulata, Prunus cerasoides, Rhus javanica, Lithocarpus dealbata, Quercus glauca, Q. griffithii, Lyonia ovalifolia, Rhododendron arboreum, Alnus nepalensis* and *Exbucklandia populnea.* The shrubby undergrowth includes *Rubus ellipticus, R. khasianus, R. rugosus, Myrsine semiserrata, Osbeckia crinita, Desmodium* sp., *Eupatorium* sp., *Lantana camara* and *Bidens pilosa.*

The tropical and subtropical evergreen and semi-evergreen forests are very rich in plant diversity and harbor large number of rare and endemic species. Contrary to these forests, pine forests are poor in the species richness. They have developed as a stable disclimax community on the disturbed subtropical broad-leaved forest sites under the influence of annual fire and other biotic disturbances. The soil profile in broad-leaved forest is well developed, acidic and rich in organic matter and nutrients. However, the soil in the pine forest is more acidic and poor in organic matter and nutrients. The tropical and subtropical broad-leaved forests found at lower and higher altitudes, respectively, represent the climax plant communities in the state. Shifting agriculture, logging, mining and other human activities have been responsible for fragmentation, destruction and degradation of climax communities giving rise to large number of secondary successional communities such as grasslands on the degraded sites.

(*v*) Grassland

In Meghalaya grasslands have developed on degraded sites after the destruction of natural forests. They occur in and around Ranikor, Mawphlang, Nongkrem, Burnihat, upper Shillong, Cherrapunji, Sutnga, Bapung, Garampani etc. areas. Dabadghao and Shankarnarayan (1973) have given an account of the grassland communities found on the hills of Meghalaya. According to them, the principal species in the grasslands at Cherrapunji are *Arundinella khasiana, Dimeria fuscescens, Ischaemum barbatum* and *Eragrostiella leioptera*, while those found near Barapani have a mixture of species from the plains and hills. *Arundinella bengalensis* is the dominant species here. At places above 1800m above sea level the grassland developed on an abandoned jhum field is dominated by *Eragrostiella leioptera* and the community is similar to that occurring at Cherrapunji. They have grouped all these grassland communities under *Themeda-Arundinella* type grass cover distributed in the entire northern and north-eastern montane tract of the country.

2.6 Forest Degradation

The primary tropical and subtropical forests including the sacred groves of the state have been destroyed/degraded to a great extent by age-old tradition of shifting agriculture, which is extensively practiced in the state even today. As a result of this and other human activities such as logging and mining, extensive degradation of the forest has taken place in the state. High rainfall and hilly terrain have further accentuated the impact of human activities on the forest. The degraded forestlands support a variety of successional communities ranging from weed-dominated communities on recently

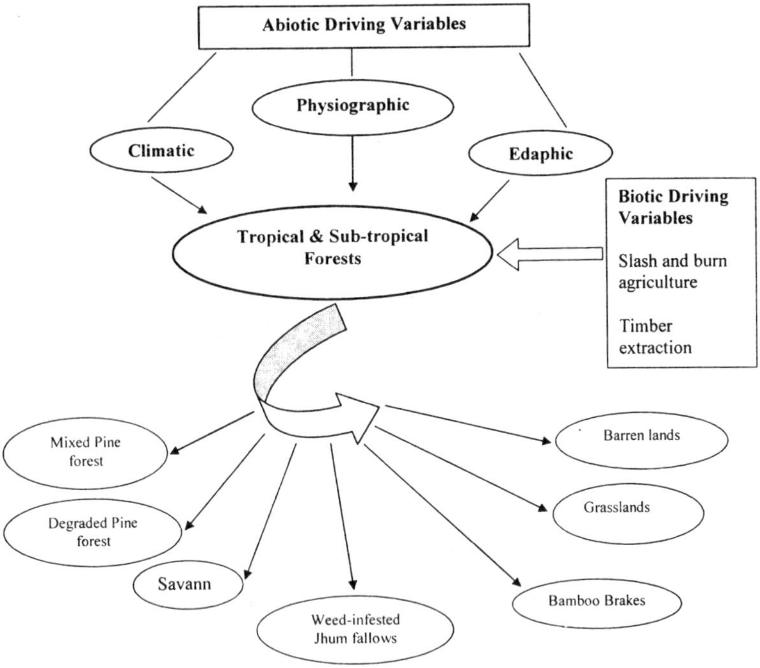

Figure 2.1: Forest Degradation in Meghalaya

abandoned Jhum fields to pine forest and grassland on frequently burnt and nutrient-deficient sites (Figure 2.1) (Pandey *et al.*, 2003).

The state with about 69 per cent of the total geographical area under forest cover is considered as a forest surplus region, but the dense forests with > 40 per cent canopy closure are getting converted into degraded open forest, scrub or grassland. At some places seismic activities, frequent landslides and resultant soil erosion have destroyed primary vegetation. While these natural causes have contributed only marginally to the change in vegetation type, it is the activity of man that has led to the irreversible transformation in the landscape.

Chapter 3

Sacred Forests of Meghalaya

The tribes of Meghalaya like other indigenous communities of India and the world also have the age-old tradition of preserving patches of old growth forests as a part of their culture and religious beliefs. These are called as *Law Kyntang* in Khasi, *Khloo Blei* in Jaintia and *Kanggimin Bol-Waarangni Biap* in Garo hills inhabited respectively by Khasi, Jaintia and Garo tribes. These forest patches have persisted through generations due to strong religious beliefs associated with them. There are large numbers of mythological stories associated with the sacred grove. The indigenous people believe that the *Guardian spirit* or *Ryngkew U Basa* rest in these forests. It is also believed that the spirits of the dead rest in peace in the *Law Kyntang*, therefore, they are considered sacred and left undisturbed. Even today a large number of undisturbed sacred forests are found in the state, though many of them have been disturbed and degraded due to human interference.

Ecologically sacred groves are the remnants of relict virgin forest vegetation of the area, which is often very much different from the vegetation of the surrounding area. They not only serve as repository of the flora and fauna of the region, but play several important roles in the life of people living in the area. Sacred groves like any other natural ecosystem perform several vital functions which are of crucial importance for the people living in its vicinity. These include purification of air and water, moderation of climate, restoration of soil fertility, production of biomass, maintenance of biodiversity, decomposition of organic wastes etc. The thick plants cover of the groves help in the conservation of water by improving rain water infiltration in the soil which in turn maintains continuous water flow in springs and streams throughout the year. A perennial source of water influences land use pattern of the locality as seen at several villages situated near the groves. The nutrients released from the decaying litter of the forest finds its way into streams which ultimately enrich the crops-fields.

Several animal species including birds that are otherwise threatened or becoming rare find safe abode in many sacred groves. A large number of animals such as rodents, amphibian, reptiles, birds, butterflies, insects and other animals depend on plants for food and shelter. At the same time many plant species in the grove depend on animal species for pollination and seed dispersal. Such species play key role in the maintenance and enrichment of the bio-diversity of the sacred groves. Over 90 per cent of the flowering plants require insects, butterflies, bees, birds, bats, and others to pollinate the flowers. A wide variety of food and micro-habitats found in the sacred groves are used by the pollinators for food and shelter that help in successful breeding and completion of their life cycles. Pollination failure has a serious consequence on existence of the species. Animals also play an important role in seed dispersal.

Many species are also valued by the local communities for cultural and religious reasons. The groves are storehouse of the ayurvedic, tribal and folk medicines. A large number of plant species found in the sacred grove have medicinal property. These species have been in use for curing large number of ailments such as gastrointestinal disorders, dermatological problems, body aches, bone fracture/dislocation, blood related problems, eye diseases, respiratory disorders and as anti-venom etc. through ages.

Sacred groves are the abode of endemic (confined to a particular area/region), rare and threatened species of the region. Khan *et al.* (1997) have reported that out of an estimated 1236 endemic species in the state, 90 are restricted to the sacred forests. Haridasan and Rao (1983) reported about 54 rare and threatened species that are mostly confined to the sacred groves in Meghalaya. In Jaintia hills 31 rare species are found in two sacred groves.

The sacred groves are also the store house of the Monoliths. In *Nong Lyngdoh*, one of the important sacred grove of Ri-Bhoi districts, there are about 178 monoliths. The highest measured monolith is of 18 feet long and is looked after by 6 clans.

3.1 Distribution of Sacred Groves

The sacred groves are found in all parts of the state and cover an approximate area of 1000 sq. km (Anonymous 1984). Tiwari *et al.* (1999) have documented eight sacred groves in East Garo Hills, eight in West Garo Hills, three in Ri-Bhoi, 32 in the East Khasi Hills, 13 in the West Khasi Hills and 15 in the Jaintia Hills district. Their size varies from 15 to 900 hectares. Subsequent to the study of Tiwari *et al.* (1999) more sacred groves were documented by Upadhaya *et al.* (2005a). These have been listed in Appendix Table 3. The list, however, is far from complete, since almost all villages in the state has one or more sacred groves. A brief description of some important ones is given in this chapter.

The vegetation of the protected groves represents the typical virgin forest of the area which varies from the tropical moist deciduous forest at lower elevation where annual rainfall is relatively less and winter temperature is relatively high. At higher elevation with relatively higher annual rainfall and low winter temperature the sacred groves represent the subtropical evergreen to semi evergreen forests.

3.2 Importance of Sacred Groves in the Life of the People

(*i*) Religious

Sacred groves are places of great religious importance for the tribal society. Annual rituals and ceremonies are performed to propitiate the deity. During these rituals, animals such as fowl, goat, pig, buffalo etc. are sacrificed. The rituals are performed for the well-being of the people, animals, crops etc. It is believed that the presiding deity takes care of the well-being of the people as well as protects the groves by administering punishment to the offenders. People take vows for wish-fulfillment at the time of difficulties; particularly during bearings of animals, birds, human, etc. At some places the sacred groves are believed to be the abodes of ancestral spirit. These groves are often used as the burial grounds. In Cherrapunji, the people believe that guardian spirit lives in the grove, and they do not enter inside it for fear of being punished by the spirit.

(*ii*) Socio-cultural

The sacred groves are the places of important socio-cultural functions. Several festivals are performed and many cultural beliefs relating to fertility and paternity find expression in vows and prayers recited at the sacred groves. People derive moral support and guidance from the cultural values associated with it. Exchange of gifts is an important social activity, which takes place at the sacred groves during certain festivals.

(*iii*) Economic

Several economic activities take place in sacred groves. There are sacred groves from were biomass is harvested and local communities derive certain direct economic benefits from them. Animal grazing, collection of dead wood, dried leaf litter and cutting of certain tree species are allowed in some groves. Spirits of sacred groves are propitiated by devotees with a view to ensure success in hunting and good harvest. Rituals are performed to bring rainfall for good health of livestock and fending off disasters.

3.3 Management of Sacred Groves

The sacred forests of Meghalaya belong either to private individuals or to the clans and communities. In the Khasi and Jaintia hills, the sacred groves are traditionally controlled by the priest called *Lyngdoh*, who is usually elected by the clan or community and holds the office till his death. A committee consisting of *Lyngdoh*, *Myntries* (ministers) and other members assist the *Lyngdoh* in framing rules for the management of the sacred forests. In Cherrapunji, *Sirdar* of the community maintains the grove. In Garo Hills, the traditional chieftain called *Nokma* is the supreme authority responsible for the protection and management of the groves. One important aspect of the management of the groves in the state of Meghalaya is that the rites and rituals are performed at regular intervals, usually once in a year by the clan under the leadership of the *Lyngdoh* in Khasi and Jaintia Hills and by the priest in the Garo hills.

3.4 Status and Threats to Sacred Groves

A survey of 56 sacred groves by Tiwari *et al.* (1999) revealed that only seven were undisturbed (100 per cent canopy), 14 were with dense canopy (>40-<100 per cent canopy), 11 were with sparse canopy (>10- <40 per cent) and 24 were with open canopy (<10 per cent canopy). Though the survey covered only a small number of sacred groves, it gives the overall picture of the status of the groves in the state. The main causes of degradation of sacred groves are encroachment for human settlement, agriculture, grazing, illicit cutting of trees etc., which is now common in the state. They are occasionally disturbed during cultural activities; for instance, straight boles of young trees of certain species are selectively taken from the grove every year during the *Beh Dein Khlam* festival in Jaintia hills for performing rituals associated with the festival. However, the greatest threat to these forests is from the age-old practice of shifting cultivation and deforestation. Some other important factors that have led to the degradation of sacred groves in the state are:

(*i*) Shift in belief system of the indigenous people from traditional animism to Christianity.

(*ii*) The most recent threats leading to degradation of sacred groves come from the process of modernization and market forces. The local traditions are being challenged by the westernized urban culture as a result the institutions of sacred groves are losing its cultural importance. The people are losing customary rights of forest management to the government.

(*iii*) The sacred groves are being fragmented by footpaths, roads, power lines, agriculture land etc.

(*iv*) Sacred groves are used for cattle grazing and illicit tree cutting particularly by the poor for their subsistence.

(*v*) Making of charcoal and hunting of animals is also causing degradation of scared groves.

3.5 Some Important Sacred Groves of Meghalaya

Raliang Sacred Groves

It is one of the most preserved sacred forests and is believed to be inherited from the deities of Khasi-Jaintia pantheon. It is worshipped by both the communities and has the status of Law *kyntang* or *Law Lyngdoh*. It is situated about 24 km north of Jowai, the district headquarters of Jaintia Hills. The Raliang 'Elaka' extends to a very large area and has 3 villages *viz.* Nongkynrih, Laskein and Raliang, each having its own headman and village council. The head of the 'Elaka', *Dolloi* resides at Raliang where 'Dorbar' administration, religious functions and other rituals are performed. The main rites and rituals are performed by the priest (*Bamon*) and a dance called *Pastieh* is performed by the men folk, accompanied by the Nongkynrih drums. The sacred groves at Raliang are located in different parts of the Raliang village. The *Blei law* sacred grove is situated on the left side of the main road leading to Garampani and covers about 0.01 ha area. The *Poh puja ka Patti* with an area of about 4 ha is also on the left side of the main road leading to Garampani. Going further eastward from the main road to Garampani, occurs a third sacred grove known as *Poh Moorang,* which has an area of about 20 ha. The fourth sacred grove *Ka pan Lyngdoh* situated on the right side of the main road, opposite to *Poh Moorang* sacred grove, covers an area of around 15 ha.

Shangpung Sacred Groves

Shangpung is situated about 74 km south-east of Shillong and about 19 km from Jowai. The 'Elaka' has 20 villages *viz.,* Shangpung, Khliehrangnah, Mynskar etc. The Dolloi of Shangpung is the head of the 'Elaka'. There are 3 sacred groves in Shangpung at different locations. The *Khlaw Byrsan* surrounded by the private forest land is situated in the south of Shangpung village with a total area of about 50 ha. The *Poh Lyngdoh* with about 30 ha area, is located by the side of the Raliang Jowai main road. The *Law Kyntang* with an area of about 400 ha is situated 3 km further south of Shangpung village. The villagers worship the guardian spirit or *U Ryngkew,* which according to their belief protects them as well as the village from natural disasters and calamities.

Khlaw Blei Sacred Grove

The sacred grove is situated at Nongthymmai Dein Shyrum village near Khleihriat and covers about 10 ha area. The village is located about 105 km east of Shillong and about 45 km west of Jowai town. The village falls under 'Elaka' Rymbai Dolloiship and consists of 50 to 60 households. At present the grove is protected and managed by the councils of *Ka Kur Lyngdoh* and *Ka Kur Sympli* clans. Cutting of trees and plants in the grove is strictly forbidden by the clan council and the offenders are punished and fined according to the norms laid down by the clan council.

Jowai Sacred Groves

Jowai town lies about 60 km east of Shillong. There are three sacred groves in Jowai: *Trepale Jowai, Khloo paiu Ram Pyrthai* and *Mokhain* covering approximately an area of 70 ha, 150 ha and 45 ha respectively. All these groves are managed by the *Sein Raij*, Jowai. The trees are not allowed to cut except for the purpose of the ceremony. The *Khloo Paiu Ram Pyrthai* sacred grove has a thick vegetation and diverse flora.

Pahampdem Sacred Grove

This sacred grove is popularly called by the local people as *U Lum Mawker* or *U Lum Umphar.* It is in Umsaw-Nongkharai Raid under the Nongpoh Sirdarship of Ri-Bhoi district. This is located about 6 km from Jyntru village, which is situated along the Guwahati-Shillong-National Highway about 55 km north of Shillong. This is the biggest (about 900 ha) sacred grove recorded in the state, but at present not more than 10 per cent of the total area is under effective forest cover. The Raid is traditionally governed by the *Lyngdoh* who is elected from the *Lyngdoh Syntiew* clan consisting of 15 villages like Pahampdem, Dulong nari, Umsaw, Nartap, Qine etc. The ceremony is performed by sacrificing one white cock and dry fish. Prawns, ginger and rice beer are offered to the deity during the ceremony.

Nonglyngdoh Sacred Grove

This grove spreads over an area of about 90 ha. It is situated at Nongkhrah village, about 55 km north of Shillong. There are 7 Raids which are governed by the head of the Raid council *i.e.*, the *Lyngdoh.* The grove is highly disturbed but the ceremony is performed like that of Pahampdem sacred grove. This grove is known for its large number of monoliths.

Cherrapunji Sacred Groves

Cherrapunji (also called *Sohra*) located at an altitude of 1300 m above msl, is about 53 km south of Shillong. There are two sacred groves in the outskirt of the town. The grove known as *Ka Law Kyntang*, with a total area of about 50 ha is near the Khliehshnong village, adjacent to the Ram Krishna Mission. The other grove *Ka Law Adong* with an area of about 900 ha is located on the left side of the main road leading to Shella. Both the groves are managed by the *Khliehshnong* 'Dorbar'.

Mawsmai Sacred Grove

Mawsmai sacred grove located about 3 km from Cherrapunji is a famous tourist spot. It is rich in cultural heritage and famous for historical background. There are 4 sacred groves: *Khlaw Ram Jadong* (150 ha), *Ka Law Blei Buh* (80 ha), *Mawlong Syiem* (120 ha) and *Ka Law Adong* (200 ha). The famous limestone caves are within the sacred grove. The *Sirdar* manages these groves.

Cherrapunji Sacred Grove

Shorarim Sacred Grove

Plate 1: Sacred Groves of East Khasi Hills, Meghalaya

Swer Sacred Grove

Mawphlang Sacred Grove

Plate 2: Sacred Groves of East Khasi Hills, Meghalaya

Khloo Paiu Ram Pyrthai Sacred Grove

Ialong Sacred Grove

Plate 3: Sacred Groves of Jaintia Hills, Meghalaya

Mawphlang Sacred Grove

Mawphlang sacred grove is one of the most well-protected and famous sacred grove in the East Khasi Hills district. It is situated near the Mawphlang village, which is about 25 km south-east of Shillong at an altitude of 1842 m. The grove has a total area of 75 ha and comprises 7 villages. The grove is looked after by the *Lyngdoh,* who is elected from four clans. Rules and regulations and management are in the hand of local 'Dorbar'. This grove is a natural treasure house of plants of the Meghalaya plateau and offers excellent scope for the professional botanists and naturalists. The vegetation of the grove is a sharp contrast from the surrounding grassland. It is also known for its monoliths erected in memory of the departed Khasi elders.

Mawnai Sacred Grove

In the West Khasi Hills, the sacred grove located at Mawnai village is a good example of the kind of sacred forests that are found in the area. Mawnai village is located at a distance of about 60 km from Shillong. It has an area of about 80 ha. The sacred grove is owned by the *Lyngdoh* of Mawnai clan and the village 'Dorbar' is held at least once in a year to discuss matters related to the protection and maintenance of the grove.

Mawlong Sacred Grove

This sacred grove with an area of about 200 ha is located in the Mawlong village, located on the left side of the main road leading to Mairang, about 41 km from Shillong. It is managed and preserved by the clan members. It is believed that the guardian spirit used to come in the form of a tiger. The *Lyngdoh* and clan members are the protector of this grove.

Nonglynkien Sacred Groves

These sacred grove falls under the jurisdiction of Maharam Syiemship and are located about 90 km west of Shillong. *U Law Lyngdoh, U Lum Sanglia* and *U Law Blei* are the three sacred groves which covers about 190 ha area. The religious rituals and ceremonies are performed every year by the *Lyngdoh* inside the grove. Strict measures have been put up for the rejuvenation of the grove.

Nonglang Sacred Grove

It is situated 5 km west of Mawkyrwat and 85 km west of Shillong. It covers an area of about 200 ha. The elders of four clan *viz., Lyngdoh, Khar Lyngdoh, Syiem* and *Lyngdoh Marshillong* clans are responsible for the management of the groves.

Mariam Sacred Grove

It is also called *Kyllai Lyngngun* and is located in Mariam village, which is about 104 km west of Mawkyrwat village. It has an area of about 80 ha. The grove is on degraded hillock with very few patches of sparsely growing trees and other plants. Formerly, the grove was under the supervision of the *Paliar* clan council, but now it is managed by the village council.

Nongbri Sacred Grove

This sacred forest is owned and controlled by the *Nongbri* clan authority of *Nongkhlaw Syiemshipare* area. The forest is located north of Pyndeng-Nongbri village about 5 km from Mairang sub-division and 50 km from Shillong. It spreads in 5 ha area. The core area of the forest is still intact and undisturbed. However, the peripheral area is degraded.

Iawen Sacred Grove

This sacred forest is situated in the village Diengsyiang near Nongkhnum River Island about 15 km from Nongstoin, the headquarters of West Khasi Hills. It covers an area of about 10 ha. The forest is controlled and managed by the *Lyngdoh Iawren* clan of Nongstoin area. The clan headquarters is in Sangriang village. Cultural rites and rituals are performed by the *Lyngdoh Iawren* clan.

Mawlot Sacred Grove

This sacred forest belongs to the *Lyngdoh Mawlot* clan of Nobosohphoh Syiemship and is located at Phyllut village. It covers an area of 20 ha. It is situated about 90 km from Shillong. The forest has been conserved in good condition by the clan authority. Traditional rites and ceremony are performed in the area regularly once in three years. Headquarter of the clan is at Marngor village.

Wanniang Sawkpoh Sacred Grove

This forest is in Shngimawlein village of Maharam Syiemship (Mawkyrwat sub-division). The area is about 7 ha. It is looked after and controlled by the *Shngimawlein* clan authority. It is situated at a distance of 22 km from Mawkrwat and 85 km from Shillong.

Diengliengbah Sacred Grove

This sacred grove is under the control of the village authority of Rngikseh, *Mawiang Syiemship* clan. The grove covers an area of 0.5 ha and is the abode of the village guardian spirit. The grove is famous in the area by the presence of the 'Birch Monumental' tree (*Betula alnoides*) which is considered to be several hundred years old. The villagers have taken steps to protect it.

Wahlang-Nongklung Sacred Grove

This sacred grove is situated at Nongklung village of Nongstoin and is controlled and managed by the *Wahlang* clan of Umiap area. It covers an area of about 10 ha. The forest area can be approached by vehicle up to Jaidoh village which is 15 km away from Nongstoin and then it takes about one hour walk by foot to reach the grove. It is a degraded grove. Many mythological stories are associated with the grove. Traditional rites and ceremony are performed by the member of the *Wahlang* clan.

Sacred Groves of Garo Hills

So far 8 sacred groves have been recorded from West Garo hills and 8 from East Garo hills. There is no report of sacred grove from South Garo hills. The grove in Garo language means forest surrounding the place of God. The area of the groves ranges from 3 ha to 37.5 ha. The Garo priest (*Sangsarik*) performs the rituals periodically. The Garos worship a number of Gods like *Tatara Rabugu, A songtatta, Songsakkia, Jaripaka etc.* All the villagers generally abide by the rules and law enacted by the headman (*Nokma*) for the protection of the groves. The offenders are punished with fines either in cash or in kind. The vegetation of the groves has evergreen and semi evergreen tree species.

Chapter 4
Floristic Diversity

The humid tropical forests of the eastern Himalayas and northeast India are floristically rich with about 7000 endemic plant species (Myers *et al.*, 2000). The floristic diversity of north-east India is unique due to its location at the confluence of the Indo-Malayan and Indo-Chinese biogeographical regions as well as its closeness to the Himalayan region and geological similarity with Peninsular India (Rao 1974). Wide variation in topography, soil characteristics and climate are the main reasons for high floristic richness of the north-eastern region which has attracted the attention of a number of botanists in the past. In the state of Meghalaya in north-eastern region, alone about 3128 species of flowering plants including 1237 endemic species has been reported by Khan *et al.* (1997).

The sacred groves of Meghalaya though located in proximity of human habitations and often small in size, retain most floristic elements of the area. They have high floristic diversity with concentration of tropical, temperate, Sino-Himalayan, Burma-Malaysian and Malayan elements. Representatives of primitive families and genera of flowering plants, large numbers of endemic and rare species, and plants of medicinal and other economic values are also found in the groves.

4.1 Floristic Composition of Sacred Groves of Jaintia Hills

A total of 547 vascular plants, both flowering and non-flowering, were recorded from the five groves (Jamir, 2000 and Upadhaya, 2002). In these groves angiosperms with 515 species were the dominant component followed by 28 species of pteridophytes and 3 species of gymnosperms (Appendix Table 4). A total of 110 angiospermic families, 15 pteridophytic and 3 gymnospermic families were identified from these groves (Pandey *et al.*, 2005)

Orchidaceae was the largest family (20–27 species) followed by Rubiaceae (17–23 species), Asteraceae (12–21 species) and Lauraceae (15–23 species) Table 4.1.

Table 4.1: Family-wise Distribution of Genera and Species in the Sacred Groves and in the Flora of Jaintia Hills, Meghalaya

Family	*Ialong Sacred Grove*		*Raliang Sacred Grove*		*Jowai* Sacred Grove*		*Jaintia Hills Flora*	
	***G*	*S*	*G*	*S*	*G*	*S*	*G*	*S*
Orchidaceae	14	20	14	22	14	27	50	141
Rubiaceae	13	17	14	19	16	23	31	75
Asteraceae	15	16	12	12	12	21	33	66
Lauraceae	7	15	10	23	6	17	9	24
Lamiaceae	9	11	5	6	10	12	20	31
Araliaceae	7	9	6	8	8	12	10	20
Rosaceae	7	11	4	8	5	11	12	29
Myrsinaceae	4	5	4	6	4	9	6	21
Moraceae	2	12	1	7	1	9	5	24
Fabaceae	8	9	5	7	6	8	25	66
Euphorbiaceae	6	9	6	6	6	8	22	44

Modified from Jamir and Pandey (2003), Pandey *et al.* (2005) and Balakrishnan (1981–83).

*: Jowai includes *Khloo Paiu Ram Pyrthai, Urkhla and Khloo Langdoh* sacred groves.

**G: Genera; S: Species.

About 81 per cent of the Orchidaceae members were epiphyte. Abundance of macro epiphytes such as orchids indicates permanently moist atmosphere, a characteristic feature of montane and subtropical as well as tropical rain forests. In closed rain forests majority of epiphytes grow high above the ground where relatively strong illumination compensates for lack of soil and the precious water supply (Richards 1996).

Presence of tropical species like *Lithocarpus fenestrata*, *Elaeocarpus floribundus*, *Sarcosperma griffithii*, *Todallia asiatica* and *Cyathea gigantae* etc., temperate species such as *Rhododendron arboreum*, *Pinus kesiya*, *Ilex* sp., *Clematis* sp. and *Ranunculus* sp. etc., and Sino-Himalayan, Burma–Malaysian species add to the floristic richness of the sacred groves. Due to presence of some species of the peninsular India region, the floristic composition of the sacred grove shows affinity with the flora of peninsular India on the one hand and provides evidence of discontinuous distribution of taxa like *Eurya japonica*, *Helicia nilagirica*, *Munronia pinnata*, *Schefflera wallichiana* etc. in India on the other (Table 4.2).

Table 4.2: Species of Different Climatic Zones and Botanical Regions Present in Sacred Groves of Jaintia Hills, Meghalaya

Tropical species	*Lithocarpus fenestrata*, *Elaeocarpus floribundus*, *Sarcosperma griffithii*, *Todallia asiatica* and *Cyathea sp.* etc.
Temperate species	*Rhododendron arboreum*, *Pinus kesiya*, *Ilex* sp., *Clematis* sp. and *Ranunculus* sp. etc
Sino-Himalayan species	*Polygala siberica*, *Corylopsis* sp., *Mahonia*, *Manglietia*, *Camellia*, *Eurya* sp., and *Anemone* sp. etc.
Burmese-Malayan species	*Vaccinium* sp., *Engelhardia spicata*, *Cinnamomum* sp., *Pittosporum* sp., *Litsea* sp., *Lasianthus* sp., *Gonionthalamus sesquipedalis*, *Balanophora dioca*, *Neillia thyrsiflora*, *Xantholis assamica* and many Zingiberaceae etc.
Peninsular-India species	*Eurya japonica*, *Helecia nilagirica*, *Munronia pinnata*, *Schefflera wallichiana* etc.

Sources: Puri (1960), Rao (1974) and Balakrishnan (1981–83).

Presence of species belonging to primitive families like Annonaceae, Rananculaceae, Piperaceae, Menispermaceae, Caryophyllaceae, Lauraceae, Myricaceae, Lazarbiadaceae and primitive genera like *Sarcandra, Corylopsis* and *Myrica* (Takhtajan 1969, Balakrishnan 1981-1983 and Rao and Hajra 1986) make the sacred groves interesting from botanical view point (Table 4.3).

Table 4.3: Primitive Flowering Plants Recorded in the Sacred Groves of Jaintia Hills

Plant Species	*Family*	*Distribution*
Manglietia sp.	Magnoliaceae	Eastern Himalayas.
Michelia sp.	Magnoliaceae	Temperate Himalayas and North-east India
Sarcandra glabra	Chloranthaceae	Indo-Malaya, North-east and southern India.
Corylopsis himalayana	Hamamelidaceae	North-east India
Myrica esculenta	Myricaceae	Himalaya, Burma, Indo-china, West and South China, Malaysia
Houttuynia cordata	Saururaceae	Himalaya, Khasi hills, Manipur, Thailand, Indo-China
Betula alnoides	Betulaceae	Temperate and subtropical Himalayas

Sources: Takhtajan (1969), Rao and Hajra (1986), Balakrishnan (1981- 83).

4.2 Floristic Composition of Sacred Groves of East Khasi Hills

Studies on floristic diversity carried out in two sacred groves of Khasi hills revealed presence of 64 tree species, 7 lianas, 22 shrubs and 25 herbs. In total 118 flowering plants including one gymnosperm (*Pinus kesiya*) were identified (Appendix Table 5). In Sohrarim sacred grove, the common tree species identified were *Daphne involucrata, Elaeocarpus lanceaefolius, Persea odoratissima, Schefflera hypoleuca, Symplocos glomerata, Syzygium tetragonum, Ficus nerifolia, Symplocos spicata, Erythroxylon kunthianum, Myrica esculenta, Magnolia pterocarpa, Castanopsis purpurella, C. tribuloides, Quercus glauca, Lithocarpus dealbata, Exbucklandia populnea, Garcinia cowa* and *Helicia nilagirica*. In Swer sacred grove, *Daphne involucrata, Elaeocarpus lanceaefolius, Persea odoratissima, Schefflera hypoleuca, Symplocos glomerata, Syzygium tetragonum, Helicia robusta, Itea chinensis, Cinnamomum bejolghota, Glochidion assamicum, Itea macrophylla, Phoebe lanceolata* and *Symplocos javanica* were the common tree species.

Among the listed species, *Daphne involucrata, Elaeocarpus lanceaefolius, Persea odoratissima, Schefflera hypoleuca, Symplocos glomerata, Myrica esculenta, Syzygium tetragonum* etc. were common to both the forest stands (Appendix Table 5).

4.3 Plant Species of Economic Importance

Out of the 435 vascular plants identified from Ialong and Raliang sacred groves in Jaintia hills, 69 species are used by the villagers for various purposes. Twenty-six species whose wild fruits are consumed by the Jaintia tribe are given in Appendix Table 6. The fruits collected from the sacred groves are not sold in the market. However, some edible fruit bearing species like *Citrus latipes, Myrica esculenta, Spondias axillaris, Pyrus pashia* are often grown in the home garden of villages. Plants used as leafy vegetables by the Jaintia tribe are listed in Appendix Table 7. Out of 16 such species some of them like *Houttuynia cordata, Polygonum dibotrys* are also cultivated in jhum fields and sold in the local markets.

Forty-five plant species recorded in the two sacred groves are used as medicine (Appendix Table 8). Several common ailments are treated by herbal medicines. For instance, eleven species are used in general body aches and colic pain, twelve species in dermatological problems, eighteen species in gastrointestinal disorders, eight species for the treatment of bone fracture/dislocation and muscular problems, three species in eye diseases, two in respiratory disorders, four in blood related problems and two as antivenin. Their use varies according to the nature and acuteness of the ailment. A given species may be used in more than one disease *e.g. Drymaria cordata* is used as antidote for snake bite and at the same time used to cure mental disorder. In gastro-intestinal disorders boiled plant extracts or plants in raw form is used. In case of dislocation of bone joints, fracture of bones and muscles sprain, pastes of plants are applied as plaster. Plants' pastes are also applied on cuts, wounds, boils and skin diseases. Even today many of the ailments are considered to be due to evil spirits and therefore religious rites are performed in order to cure these diseases.It was noticed while interacting with the people of the area that the young folks have little interest/knowledge about the local medicinal plants and therefore it is very likely that the traditional knowledge about medicinal plants will slowly disappear in course of time.

4.4 Rare, Endemic and Threatened Plant Species

Ninety-one species recorded in five sacred groves of Jaintia hills in Meghalaya are either rare, endangered in Meghalaya or endemic to northeast India or Meghalaya (Appendix Table 9). Out of 91 species, 60 species are endemic to northeast India or eastern Himalayas and 51 are rare to Meghalaya. The distribution of these species shows that 88 species are found in the Indian subcontinent, 45 species are exclusive to northeast India and 26 species are endemic to Meghalaya. The species endemic to Meghalaya includes *Aeschynanthes parasiticus, A. superba, Callicarpa psilocalyx, Camellia caduca, Citrus latipes, Ilex embeloides, Impatiens khasiana, I. laevigatum, Lindera latifolia, Paramignya micrantha* and *Rubus khasianus* (Balakrishnan 1981-83). According to Takhtajan (1988) the flora of the Khasi and Jaintia hills is most richly saturated by eastern Asiatic elements, and the area is one of the most important centers of survival of the tertiary flora of eastern Asia. The geological antiquity of the Khasi hills, the taxonomic composition of the flora and its richness and peculiarity indicates that long before glacial times, these mountains constituted an important part of the area where the nucleus of the flora of eastern Asiatic flora originated (Takhtajan 1988). Concentration of endemic species in the groves reveals that even today they are the abode of the endemic taxa of the region.

A list of 51 species based on their rarity status in Khasi hills of Meghalaya is given in Appendix Table 10. Two rare ferns namely, *Dipteris wallichii* and *Cyathea gigantea* and *Ilex embeloides, Styrax hookerii* and *Fissistigma verrucosum,* were collected after several decades from the sacred groves of Jaintia hills (Jamir 2000, Upadhaya 2002). Some orchid species such as *Dendrobium* and *Pleione* having ornamental value are becoming rare in nature due to their over-exploitation, while *Podocarpus neriifolia, Cyathea gigantea* and *Balanophora dioca,* are becoming rare due to habitat destruction. Rao and Haridasan (1983) have reported 54 rare and threatened plants and Haridasan and Rao (1985-1987) have listed 44 rare dicotyledonous plants from Meghalaya. Compared to these figures the sacred groves have high concentration of rare plant species. Population size of such species in the groves is small due to combination of unfavourable regeneration conditions, lack of appropriate habitats or both.

Plate 4: Endemic and Rare Plant Species Found in Sacred Groves of Meghalaya

1. *Fissistigma verrucosum,* 2. *Aeschynanthes sikkimensis,* 3. *Aeschynanthes superba,* 4. *Dipteris wallichii,* 5. *Balanophora dioca,* 6. *Ilex khasiana* (1, 2, 3 are Endemic to north-east India and 4, 5 are Rare to Meghalaya and 6 is Endemic to Meghalaya)

Contd...

Plate 4–Contd...

3

4

Contd...

Plate 4–Contd...

5

6

The rich floristic diversity of the groves makes them invaluable natural resource and call for immediate steps for their protection to conserve rich plant diversity of the region. These forest patches show concentration of tropical, temperate, Sino-Himalayan, Burma-Malaysian and Malayan elements as well as representatives of primitive families of flowering plants. The high concentration of endemic and rare species in these sacred groves is attributed to their geographical proximity to the species-rich Eastern Himalayas, South Central China, Burma and Malaya. The favourable climatic conditions of the area and prolonged protection by the local tribal people have also played a key role in making the sacred grove species-rich. The species richness was not uniformly distributed in the groves; rather they were mosaic of low and high diversity patches. In this respect they are similar to tropical rainforests, which have described as highly patchy communities (Ashton, 1969; Poore, 1968). However, due to increasing anthropogenic pressure on land and with the gradual erosion of traditional beliefs, the groves are being disturbed in various ways by the local tribes, and they themselves are threatening the existence of these species-rich remnant plant communities of the past vegetation.

Chapter 5

Community Structure: Life-form, Stratification and Tree Population Structure

The external appearance or physiognomy of sacred grove communities found at different altitudes in the state varied with the change in the vegetàtion type from the moist deciduous forest at lower altitude to semi-evergreen to evergreen forests at the higher altitude. In general, the average height of the canopy trees in the sacred groves was 18m. The height of some tree species such as *Ficus* sp., *Betula alnoides, Engelhardtia spicata, Acer laevigatum, A. oblongum* often reached about 25m in most of the studied sacred groves. However, in *Urkhla* sacred grove in Jaintia hills, *Betula alnoides* with a height of > 33m was the tallest tree. In terms of height of canopy trees and other tree characteristics the sacred grove communities resembled with the subtropical evergreen broad leaved forests of South China (Box 1995, Song 1995, Coa and Rob 1997). A comparison of physiognomic characteristics of the lowland and montane tropical forests presented in Table 5.1 shows similarity in tree height, principal leaf size class of woody species and epiphytes of the sacred grove vegetation with that of the tropical lower montane forest communities.

Table 5.1: Comparison of Tree Characteristics of Sacred Groves with Other Tropical Forests

Parameters	*Lowland Evergreen Rainforests*	*Lower Montane Forest*	*Upper Montane Forest*	*Sacred Groves*
Canopy height	25-45m	15-33m	1.5-18m	15-30m
Emergent trees	Present, 60-80m tall	Often absent to 37m tall	Usually absent to 26m tall	Usually absent to 30m tall
Principal leaf size of woody plants	Mesophyll	Mesophyll	Microphyll	Mesophyll
Buttreess	Usually frequent and large	Uncommon, small	Usually absent	Uncommon, small
Big woody climbers	Absent	Usually none	None	Frequent
Bole climbers	Often abundant	Frequent to abundant	Very few	Frequent to abundant
Vascular epiphytes	Frequent	Frequent to abundant	Frequent	Abundant
Non-vascular epiphytes	Occasional	Occasional to abundant	Often abundant	Abundant

Sources: Raunkiaer (1934), Whitmore (1975), Jamir (2000).

5.1 Life-form Spectrum

Several quantitative measures have been employed to describe the structure of plant communities with ecological precision. Among them species richness, life-form spectrum, vertical disposition of species, and their phenology are important parameters to describe community structure. Raunkiaer (1934) described communities of different climatic zones or phytoclimatic zones of the earth on the basis of life-form spectrum. Any change in the life-form composition away from its phytoclimatic zone is considered as an indicator of alteration in vegetation either due to climatic or edaphic factor or both. However, in recent years man has abused the natural vegetation surrounding them so much so that their life-form composition is often determined more by the biotic stresses than by the climatic and edaphic factors.

Life-form composition of five sacred groves of Jaintia hills of Meghalaya were investigated by Jamir (2000) and Upadhaya (2002). In all of them the phanerophytes were the dominant (44 per cent–51 per cent) life-form. The percentage of therophytes ranged from 2 per cent to 12 per cent, the epiphytes contributed 11 per cent to 16 per cent, and lianas and climbers together constituted about 7 per cent to 18 per cent in the community (Figure 5.1).

Dominance of phanerophytes and high concentration of lianas and climbers brings the sacred groves closer to the tropical moist and humid forests (Daniels *et al.*, 1995, Vazquezg *et al.*, 1998). The high percentage of epiphytes in the community is attributed to high precipitation and high atmospheric humidity all the year round. Such a climatic condition is generally found in the mid-montane forests (Benzing 1983). In the tropical belt, places with increased aridity, deforestation and other anthropogenic disturbances often result in the dominance of therophytic or thero–chamaephytic conditions. Presence of chamaephytes and therophytes in sacred groves was the result of human disturbance of mild intensity.

5.2 Stratification

A mature forest is a stratified plant community where plants of different life-forms occupy different vertical space starting from the forest floor to the canopy. The number of stratatum, however, varies from one forest type to another depending primarily on the prevailing climatic condition. For instance a low land humid tropical forest is

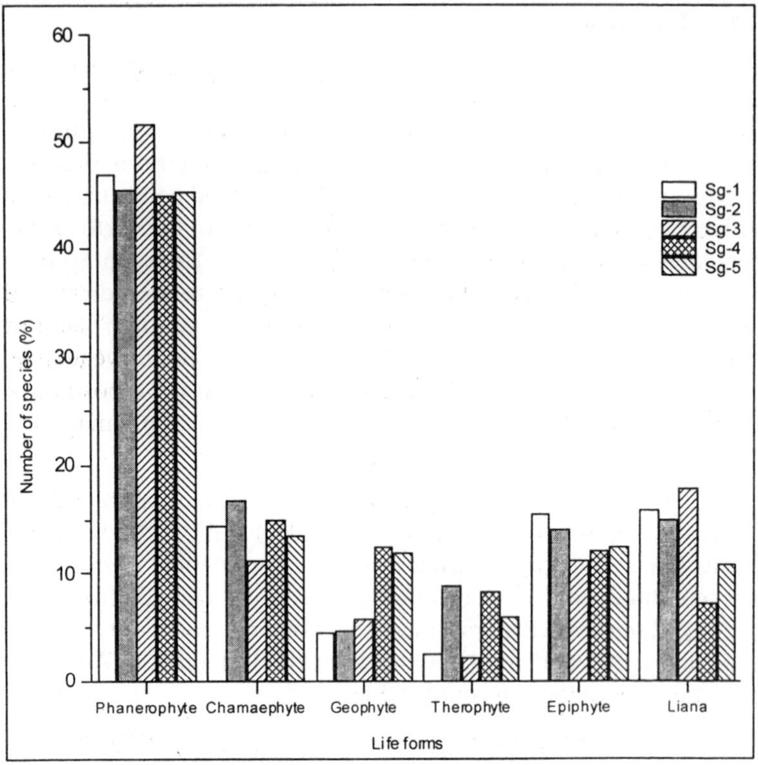

Figure 5.1: Life-form Composition of Sacred Grove Communities of Jaintia Hills

a highly stratified community while a temperate or alpine coniferous forest has virtually one layer of the plants. Lianas also play an important role in structuring canopy architecture of the forests. Recently quantitative data and aerial photography have been widely employed to understand the vertical structure of the forest, but in most instances it has failed to correctly portray the forest stratification. Profile diagrams of sample strips of forest to the scale are useful in this respect (Richards 1996). This technique was first applied in the forest of British Guyana by Davis and Richards (1933-34). Since then it has been widely used in many parts of the world. Vegetation profile of the component species indicate the phytoclimatic condition of the area and are useful in depicting the general features of the forest structure that are not easily appreciated from photographs and quantitative data alone. They are particularly valuable in comparing forest types differing in structure.

Species richness in different growth forms and profile of five sacred grove communities of Meghalaya has been prepared by Jamir (2000) and Jamir *et al.* (2006). In all the groves tree layer was most species rich followed by herb layer, understorey shrubs/epiphytes, liana/climbers/twiners and scandent shrubs/stragglers (Table 5.2).

Table 5.2: Distribution of Species in Different Synusiae of the Sacred Groves of Jaintia Hills

Growth Forms	*Sacred Groves*				
	Sg1	*Sg2*	*Sg3*	*Sg4*	*Sg5*
Trees	106 (31)	110 (30)	97 (39)	122 (32)	98 (48)
Scandent shrubs/ stragglers	20 (6)	23 (6)	15 (6)	16 (5)	17 (8)
Liana/climbers/ Twiners	26 (7)	33 (7)	28 (10)	34 (10)	23 (10)
Epiphytes	33 (10)	38 (9)	41 (15)	47 (14)	25 (11)
Shrubs	49 (14)	44(14)	30 (11)	30 (9)	18 (8)
Herbs	113 (33)	102(32)	57 (21)	85 (25)	42 (19)
Total	347	350	269	334	223

Values in the parentheses are percentage of the total number of species.

The vegetation profiles of *Khloo Paiu Ram Pyrthai* (Sg–1), *Urkhla* (Sg–2), *Khloo Langdoh* (Sg–3), *Khloo Blai* (Sg–4) and *Khloo Poh-Lyngdoh* (Sg–5) sacred groves drawn in belt transects (7.5m wide and 80–120m long) are depicted in Figure 5.2, 5.3, 5.4, 5.5, and 5.6 respectively.

In *Khloo Paiu Ram Pyrthai* there were 45 individuals distributed in three distinct layers *viz.*, canopy, sub-canopy and under canopy layers within a 90m long and 7.5m wide transect. The discontinuous canopy layer having a height of 20 m was composed of *Schima khasiana, Cinnamomum glanduliferum, Castanopsis tribuloides, Persea odoratissima* and *Fraxinus floribunda* species. The canopy layer was interrupted at places by gaps that were filled by trees of the sub-canopy layer. The common species in this layer were *Helicia nilagerica, Symplocos cochinchinensis, Xantolis assamica, Turpina nepalensis,* and *Vaccinium sprengelii*. Young trees of the canopy layer such as *Schima khasiana, Rhus acuminata,* were also found in this layer. Below the sub-canopy, there was a dense layer in which *Coffea khasiana, Styrax serrulatum,* and *Camellia caudata* were conspicuous species.

In the profile diagram of *Urkhla* forest (Figure 5.3) *Cinnamomum glanduliferum, Castanopsis tribulioides, Betula alnoides, Syzygium tetragonum* were common species in the canopy layer. *Betula alnoides* was the tallest tree (average height >> 33m) of this layer. The subcanopy was composed of *Quercus kamroopii, Helicia nilagirica, Lithocarpus elegans, Croton oblongus, Pithecellobium monadelphum,* and *Myrsine semiserrata. Erythroxylum kunthianum, Daphne involucrata, Camellia caudata* were common under canopy species.

In *Khloo Langdoh* the canopy and sub-canopy layers were indistinct; both formed a single undulating layer in which tree height was about 20m (Figure 5.4). The species present in this layer were *Drimycarpus racemosus, Acer laevigatum, Cinnamomum bejholghota, Garcinia anomala,* and *Glochidion lanceolarium*. The under canopy was composed of *Coffea khasiana, Microtropis discolor* and *Acronychia pedunculata* species etc.

The *Khloo Blai* (Figure 5.5) forest had uninterrupted canopy cover. The forest near hill top had two layers of trees, while that on the slope had three layers. The canopy layer growing to a height of 25m was composed of *Castanopsis tribuloides, Castanopsis purpurella, Betula alnoides, Cinnamomum glanduliferum* and *Alseodaphne petiolaris.* Young individuals of canopy species such as *Castanopsis purpurella, Alseodaphne petiolaris, Schima wallichii, Rhus acuminata, Alangium chinense, Ficus cocinna* etc. were present in the sub-canopy. The under canopy layer was quite dense. The common species in this layer were *Camellia caudata, Eurya acuminata, Ficus nerifolia, Itea macrophylla* etc.

Khloo Poh Lyngdoh sacred grove (Figure 5.6) showed three distinct layers. The canopy layer was composed of *Engelhardtia spicata, Castanopsis purpurella, Sarcosperma griffithii, Acer laevigatum* etc.; the sub-canopy layer was constituted by *Prunus jenkensii, Sarcosperma griffithii, Knema angustifolia, Pithecellobium monadelphum* etc. Some scandent plants like *Fissistigma verrucosum* grew to a height of canopy layer. The under canopy layer was dominated by *Capparis acutifolia, Lindera latifolia, Citrus latipes* etc.

In two well protected sacred groves at Swer and Shorarim in the East Khasi hills the trees were distributed in three distinct layers

Figure 5.2: Profile Diagram of Khloo Paiu Ram Pyrthai (Sg-1): ***Schfellera hypoleuca*- 1, *Schima khasiana*- 2,5, *Vaccinium speringelii*-3, 40, *Persea odoratissima*- 4, 10, 38, *Castanopsis tribuloides*- 6, 7, 9, 18, *Saurauia nepalensis*- 8, *Styrax cerrulatum*- 11, *Pithecellobium monodelphum*- 12, 16, 28, *Camellia caudate*- 17, *Helicia nilagirica*- 13, 32, 37, *Coffea khasiana*- 14, 13, 39, 42, *Cinnamomum glanduliferum*- 29, 31, 44, *Rhus acuminate*- 23, 34, *Symplocos cochinchinensis*- 15, 19, 36, *Ficus hirta*- 21, *Xantholis assamica*- 27, 35, *Engelhardtia spicata*- 22, 25, 26, *Myrsine semiserrata*- 24, *Microtropis discolor*- 33, *Lithocarpus fenestrate*- 41, *Turpina nepalensis*- 43, *Fraxinus floribunda*-45.**

Figure 5.3: Profile Diagram of Urkhla (Sg-2): ***Persea parviflora*-** **1, 4, 52,** ***Pithecellobium monodelphum*-** **2, 42, 66,** ***Castanopsis tribuloides*-** **3, 13, 27, 38, 51, 57,** ***Coffea khasiana*-** **5, 11, 68,** ***Persea globularia*-** **7, 10,31,** ***Camellia caudate*-** **6, 8, 45, 46, 49, 50,** ***Quercus kamroopii*-** **9, 19, 29, 63,** ***Helicia nilagirica*-** **12, 14, 16, 17, 20, 41, 56,** ***Syzygium tetragonum*-** **15, 18, 23, 37, 43, 55, 60, 72,** ***Cinnamomum glanduliferum*-** **25, 26, 53, 62,** ***Myrsine semiserrata*-** **21, 22, 36, 44, 71,** ***Styrax serrulatum*** **-24, 32, 40, 58, 70,** ***Erythroxylon kunthianum*-** **28, 33, 48,** ***Quercus kamropii*-** **30,** ***Symplocos cochinchinensis*-** **34, 35,** ***Croton oblongus*-** **39,** ***Betula alnoides*-** **47,** ***Lithocarpus elagans*-** **54, 64, 69, 74,** ***Cinnamomum parviflorum*-** **61,** ***Daphne involucrate*-** **65, 67,** ***Tetrastigma leucostaphyllum*-** **73.**

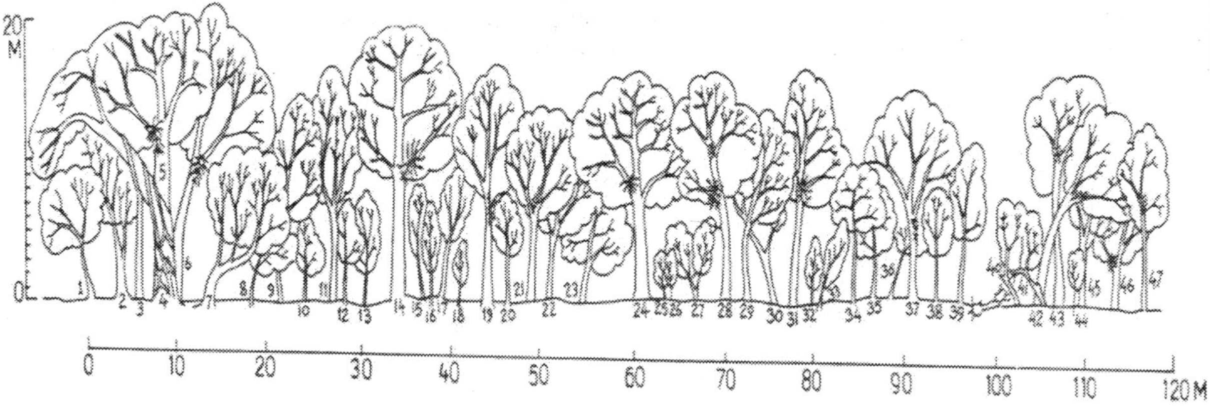

Figure 5.4: Profile Diagram of Khloo Langdoh (Sg-3): ***Pithecellobium monadelphum*** **- 1, 7, 13,** ***Xantholis assamica*** **- 2,16, 17,** ***Acer laevigatum*** **- 3,8,24,** ***Cinnamomum bejolghota*** **- 4, 36,** ***Ficus concinna*** **- 5,** ***Drimycarpus racemosus*** **- 6, 31, 43,** ***Glochidin lanceolarium*** **- 9, 22, 30, 39,** ***Styrax serrulatum*** **- 10, 32,** ***Castanopsis tribuloides*** **- 11, 14, 19,** ***Coffea khasiana*** **- 12, 15, 25, 26, 44,** ***Camellia caudata*** **- 18,** ***Persea odoratissima*** **- 20, 37, 47,** ***Garcinia anomala*** **- 21, 29, 46,** ***Lindera latifolia*** **- 35, 41,** ***Evodia trichotoma*** **- 23,** ***Castanopsis purpurella*** **- 28, 42,** ***Microtropis discolor*** **- 27, 33,** ***Acronychia pedunculata*** **- 34, 38,** ***Helicia nilagirica*** **- 40, 45.**

Figure 5.5: Profile Diagram of Khloo Blei (Sg-4): ***Castanopsis tribuloides*-** **1, 3, 38,** ***Schima wallichii*-** **2, 4,** ***Rhus acuminate*-** **5, 12, 19,** ***Ficus nerifolia*-** **6, 10,** ***Betula alnoides*-** **7,** ***Eurya acuminate*-** **8, 24,** ***Alangium chinense*-** **9, 13,** ***Camellia caudata*-** **14, 15, 26, 27, 31, 33, 34, 35, 39, 43, 46,** ***Engelhadtia spicata*-** **17,** ***Castanopsis purpurella*-** **11, 16, 18, 25, 36, 40,** ***Pyrularia edulies*-** **20, 29,** ***Alseodaphnae petiolaris*-** **21, 22, 23,** ***Ficus concinna*-** **28,** ***Cinnamomum glanduliferum*-** **30, 32,** ***Eurya cerasifolia*-** **37,** ***Symplocos spicata*-** **41,** ***Coffea khasiana*-** **42,** ***Phobe lanceolata*-** **44,** ***Itea macrophylla*-** **45, 48,** ***Meyna spinosa*-** **47.**

Figure 5.6: Profile Diagram of Khloo Poh-Lyngdoh (Sg-5): *Engelhardtia spicata*-1, *Macropanax dispermus*- 2, 9, *Sarcosperma griffithii*- 3, 5, 7, 16, 18, 24, 33, 40, *Castanopsis purpurella*- 4, 8, 12, *Actinodaphne obovata*- 6, 15, *Capparis acutifolia*- 10, 22, 25, *Prunus jenkensii*- 11, 23, 30, 31, 35, 36, *Fissistigma verrucosum*- 13, 17, *Garcinia tinctoria*- 14, 38, *Acer laevigatum*- 19, 29, *Tupidanthus calyptratus*- 20, *Dysoxylon gobara*- 21, 27, *Neolitsea cassia*- 26, *Spondias axillaries*- 28, *Citrus latipes*- 32, *Lindera latifolia*- 34, *Pithecellobium monadelphum*- 37, *Acer oblongum*- 39, *Knema angustifolia*-41, 42, Unidentified species- 43, *Ficus* sp.- 44.

i.e., canopy layer >15m, sub-canopy 8-15m and under canopy <8m. In Sohrarim, *Castanopsis purpurella, Castanopsis tribuloides, Lithocarpus dealbatas* and *Quercus glauca* were common canopy species. *Schefflera elata, Symplocos* sp, *Photinia integrifolia, Ilex venulosa, Helicia robusta, Ilex embeloides* and *Daphne involucrata* were common species identified in the sub-canopy layer. Species like *Camellia caduca, Camellia drupifera, Daphne papyracea, Pentapanax subcordatum, Symplocos spicata* and *Symplocos pyrifolia* were common species in the under canopy.

In Swer canopy species included *Myrica esculenta, Rhododendron arboreum, Magnolia pterocarpa, Persea odoratissima* and *Daphniphyllum himalayense.* In the sub-canopy *Daphne involucrata, Ilex mbeloides, Ilex theaefolia, Ilex venulosa, Litsea salicifolia* and *Phoebe lanceolata* were common species. In the under-canopy *Camellia caduca, Itea chinensis, Daphne papyracea, Symplocos pyrifolia* and *Symplocos spicata* were dominant species.

The sacred groves showed similarity in species composition with the subtropical evergreen broad leaved forest of South China located at similar latitude (24°-32°N) with preponderance of Fagaceae, Lauraceae, Theaceae etc. Presence of species of diverse growth forms and well developed stratification with an average tree height in the over storey ranging between 15m and 25m, rarely crossing above 30 m, is another feature similar to the forest found in southwest China (Song 1995, Box 1995, Cao and Rob 1997).

The tropical rain forest, which is considered the most complex and highly organized terrestrial community in the world, has five or six distinct strata (Richards 1996). In most of them, the emergent trees constitute the top layer or A stratum and the B and C strata form a continuous canopy layer; often with indistinct boundary between the two (Ashton and Hall 1992, Pascal 1992, Richards 1996). In the sacred groves emergent trees were absent and the average tree height in the canopy layer is comparable to the height of the B stratum of the humid tropical forests at Moraballi Creek and Sarawak (Richards 1996), Brunei (Ashton and Hall 1992) and Brazil (Thompson *et al.*, 1992). The height of canopy trees is influenced by several factors. For instance, forests developed in dry climate and on unproductive soil have shorter canopy height (Whitmore 1975, Sahunalu and Dhanmanonda 1995 and Visalakshi 1995). Since Meghalaya receives plenty of rainfall, aridity does not appear to be the main cause for shorter tree height. The soils developed under high rainfall condition in the tropics are highly leached and poor in nutrients. This could be one of the reasons for relatively shorter tree heights in the sacred groves, besides higher elevation which also influences tree height (Whitmore 1975, Grubb 1977, Lieberman *et al.*, 1996, Aiba and Kitayama 1999). However, shorter height of the canopy trees and indistinct canopy and sub-canopy strata found in *Khloo Langdoh* and upper ridges of *Khloo Blei* sacred groves could also be attributed to the selective extraction of tree species with straight bole for the annual *'Beh Dein Khlam'* festival, and other anthropogenic disturbances.

5.3 Tree Population Structure

Distribution of individuals of different age groups in a population reflects its history of survival, reproduction and potential for future growth. Age distribution indicates periods of successful reproduction, periods of high and low survival, and whether the older individuals in a population are replacing themselves or if the population is declining. Inventories of natural population of tropical trees by girth or diameter classes are numerous. In natural rain forests there are many more young individuals than mature trees in the majority of species and in the population as a whole (Richards 1996). The diameter class distribution also reflects the height distribution to some extent and gives some indication of the vertical structure of the forest and also reflects the regeneration potential of the forests. Jamir (2000) and Upadhaya *et al.* (2004) studied the tree population structure in few sacred groves of Jaintia hills in Meghalaya and reported preponderance of trees in the lower dbh (5-15 cm) class.

An analysis of distribution pattern of individuals of different age groups starting from seedlings to mature trees showed an upright pyramid with the seedlings forming the base and mature trees occupying its top (Figure 5.7). This type of tree population structure is a characteristic feature of the forest that develops on moist and infertile soil and has also been reported from mature forests of Brazalian Amazon (Campbell *et al.*, 1992), Costa Rica (Nadkarni *et al.*, 1995) and Eastern Ghats (Kadavul and Parthasarathy 1999). It is interesting to note that though the population structure shows the characteristics of an expanding population, the forest hardly shows any sign of lateral expansion. This clearly shows that though the microclimatic conditions in the canopy gaps and forest understorey favoured germination and tree seedlings recruitment, but they do

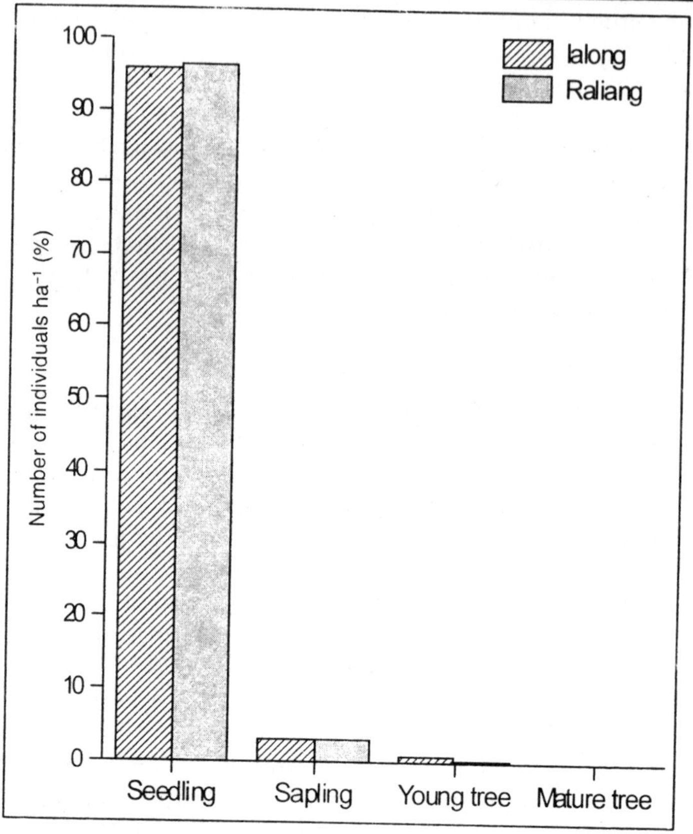

Figure 5.7: Population Structure of Woody Species in Ialong and Raliang Sacred Groves

not survive for long for a variety of reasons that include moisture stress during post monsoon period, low temperature during winter and competition for light, herbivory etc. The surviving seedlings have to face several abiotic and biotic stresses which hamper their growth; light being the most important factor in this respect. Under low light condition they grow at a very slow pace till light condition again improves after creation of gaps in the over head canopy either by natural agencies or by anthropogenic activities such as selective removal of trees. All these factors together were responsible for preponderance of individuals of lower girth classes in the sacred forests.

In most cases sacred groves are found as isolated patches surrounded by denuded or degraded areas. The soil condition and microclimate outside the forest are vastly different from that found within the forest. This prohibits germination and seedling establishment of majority of species which have very specific environmental requirements and require specialized niches for their establishment. Therefore in majority of cases sacred groves do not show any lateral expansion. This is particularly true for smaller groves where large peripheral area is exposed.

Chapter 6

Community Structure: Species Dispersion, Diversity and Dominance

The structure of a plant community is described on the basis of large number of qualitative attributes and quantitative parameters of the species that constitute it. Among the quantitative parameters, dispersion pattern, diversity and dominance of species are important determinants of the community structure. A number of formulae, indices and models have been devised by ecologists to explain various aspects of species richness in a community. The most widely and accepted tool for measuring diversity is the Shannon and Weiner's (1963) index, commonly known as Shannon's Index of general diversity. Simpson's (1949) successfully formulated an index known as Simpson's index of dominance. This index shows a high preponderance of the most abundant species in the sample. Thus the value of Simpson's index increases as the index of general diversity decreases (Magurran 1988). Whittaker (1972, 1975) used α (alpha) and β (beta) diversity to measure species richness of a community or habitat and diversity between habitats or communities respectively. Jaccard (1912) and Sorensen (1948) derived formulae to measure similarity between habitats or communities based on species composition. These are qualitative in nature. The quantitative approach to the measurement of similarity between the habitats or communities was worked out by Morisita-Horn (Morisita 1959, Horn 1966 and Magurran 1988).

A number of studies on plant diversity have been carried out in the tropics and subtropics. Gentry (1988), Felfili (1995), Johnston and Gillman (1995), Lieberman *et al.* (1996) are some of the prominent workers who made significant contribution to the study of plant diversity. Hubbell and Foster (1983) carried out an exhaustive survey of 50 ha permanent plot in Borro Colarado Island, Panama. Similarly a 50 ha plot was surveyed by Manokaran *et al.* (1991) and Sukumar *et al.* (1992) in Malayasia and South India respectively. Turner and Corlett (1996), Aiba Shin-ichiro and Kitayama (1999), Cao and Rob (1997) and Sahunula and Dhanmononda (1995) have contributed to the understanding of biodiversity in the tropical and subtropical Asia.

In India, Parthasarathy and Karthikeyan (1997), Sukumar *et al.* (1992), Visalakshi (1995), Ganesh *et al.* (1996), Joshi *et al.* (1997), Ganeshieh *et al.* (1997), Kadavul and Parthasarathy (1999), Jamir (2000), Upadhaya *et al.* (2003, 2004) have contributed to the understanding of the plant diversity in different parts of the country. Ayyapan and Parthasarathy (1999) studied 300 ha permanent plot in Western Ghats of India.

6.1 Species-area Curve

A quantitative relationship between area and number of species has been studied on islands, habitat patches on continents and along latitudinal gradient by a number of workers. Their findings suggest that species richness on islands and habitat patches like contracted vegetation on mountain tops and lakes increases with area and decreases with isolation. Such a relationship between area and species is considered as the outcome of a highly dynamic balance between immigration and extinction of species which is mainly determined by the isolation and area of the islands.

The sacred groves being isolated patches of vegetation surrounded by contrasting vegetation like grasslands, crop fields, denuded habitats, human habitation are also like an island so far as the area and number of species is concerned. Species-area relationship in three sacred groves depicted in Figure 6.1 showed a gradual increase in the tree species number with increase in the area (Jamir 2000). However, the number of species did not increase beyond

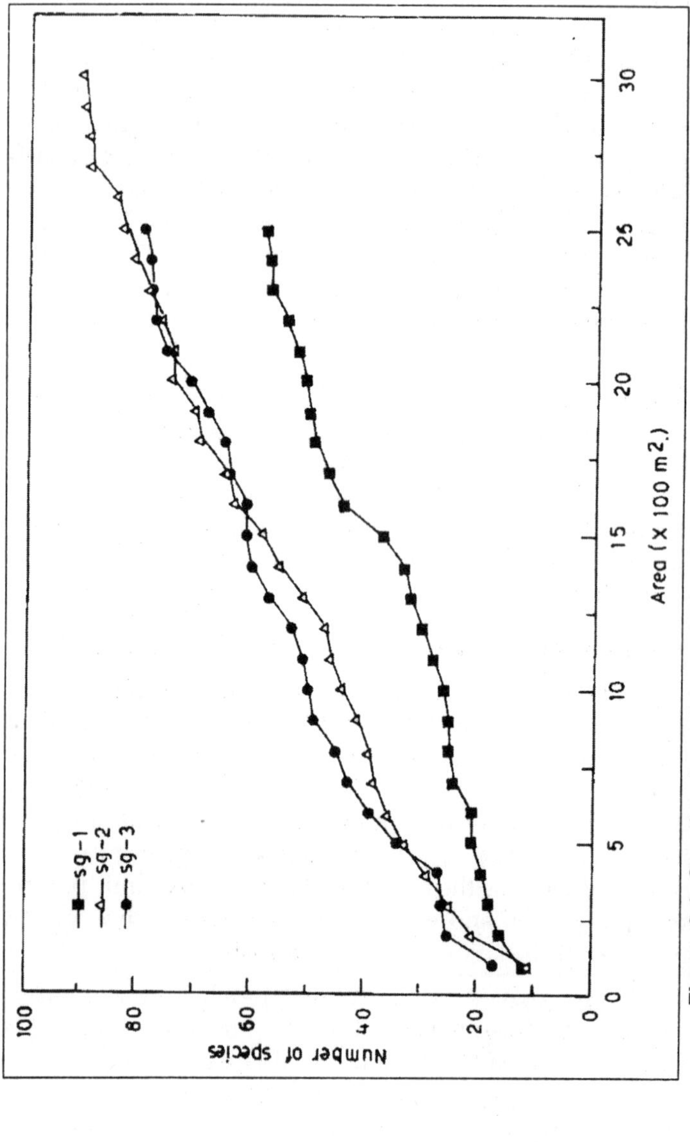

Figure 6.1: Species Area Curve of Tree Species in Sacred Groves of Jaintia Hills (Sg-1 Khloo Paiu Ram Pyrthai, Sg-2 Urkhla, Sg-3 Khloo Langdoh) (after Jamir 2000)

0.35–0.4 ha area. In each grove about 80 per cent of the total species encountered were found in 0.35 ha area, while 88–99 per cent of the species were present in 0.4 ha area in all the studied sacred groves.

6.2 Dispersion Pattern

In communities' plants are not distributed uniformly in the horizontal space, rather they may be dispersed in random, contagious or regular fashion, expressed in the patchy or mosaic occurrence of species in the communities. Communities also show temporal differentiation, with different species carrying on similar functions at different times in seasonal and daily cycles. In a given community each species has its own place in vertical and horizontal space, time and way of relating to other species. The species' place in the community in relation to other species is called its niche (Whittaker 1975). The dispersion patterns of tree species in horizontal space not only influence the overall structure of the forest, it also reveals species niche in the community. In all rain forests, the trees are unevenly, but usually not randomly spaced. This applies to individuals of one species as well as the tree population as a whole and seems to be due to various reasons, some intrinsic to the trees themselves and some influenced by environmental condition. Regular patterns are rare and are found sometimes in peat swamp forests. Clumping of individuals of the some species may be due to opportunity or chance and is often clearly related to gap formation and dispersal mechanism of the species. When numerous saplings are able to grow in large gaps; it is often clearly related to the dispersal mechanisms of the species. Armesto *et al.* (1986) compared the dispersion pattern of trees in tropical and temperate forests in different parts of the world and concluded that clumping was characteristic of forest in which formation of canopy gaps was the chief source of disturbance. Hubbell (1979) in dry tropical forest observed that all species were either clumped or randomly dispersed, with rare species more clumped than common species.

Dispersion pattern of tree species studied two sacred groves revealed that majority of the species showed clumped distribution pattern both in protected and disturbed groves. This clearly suggests that canopy gaps play an important role in determining the distribution of species on the forest floor and making the forest highly heterogeneous and patchy in terms of species distribution in horizontal space (Table 6.1).

Table 6.1: Distribution Pattern of Woody Species in Two Sacred Groves of Meghalaya (After Upadhaya *et al.*, 2004)

Sacred Groves	*Distribution Pattern (Number of Species)*		
	Regular	*Random*	*Contagious*
Ialong			
Undisturbed	0	8 (10)	74 (90)
Disturbed	1 (1)	5 (6)	74 (93)
Raliang			
Undisturbed	0	5 (5)	87 (95)
Disturbed	0	7 (8)	86 (92)

The number in parentheses is the percentage of the total species.

6.3 Species Diversity

Several hypotheses have been put forward to explain the species diversity in communities. These may be broadly categorized in to two groups. One group of theories (equilibrium theories) considers the environment as stable and recognizes intra- and inter-specific competition as the determinants of species diversity. The importance of competitive abilities of the species for co-existence has been stressed in this theory. It is argued that species though survive in a particular environment; they are able to avoid or reduce competition from other species by finding its own niche. Another group of theories (non-equilibrium theory) focuses on the spatial and temporal instability of the environment which creates regeneration niches for species (Grubb 1977).

It has been argued by several workers (cf. Brown 1981) that productivity of the system and structural complexity or heterogeneity determines species richness in the community. Putman (1994) argued that neither productivity nor structural complexities seem sufficient on their own to explain the observed pattern of species richness. Slobodkin and Sanders (1969) opined that species richness of any community is a function of severity, variability and predictability of environment in which it develops.Therefore diversity is seen to increase as environment become more favourable and more predictable (Putman 1994).

Table 6.2: Species Diversity in Sacred Groves of Northeast India

Sacred Grove	*Shannon Diversity Index*	*Evenness Index*	*Simpson Dominance Index*	*Authors*
Meghalaya Sacred Groves				Upadhaya *et al.*,
Raliang (Jaintia hills):				2004, 2006
Woody species (≥ 5cm dbh)	3.55	0.56	0.034	
Shrubs	3.04	0.55	0.078	
Herbs	3.30	0.60	0.049	
Ialong (Jaintia hills):				
Woody species (≥ 5 cm dbh)	3.42	0.53	0.067	
Shrubs	3.40	0.61	0.048	
Herbs	3.26	0.57	0.067	
Swer (East Khasi hills)	2.2-2.8	0.40	0.10-12	Mishra *et al.*, 2004 Upadhaya *et al.*, 2008
Nonglang (East Khasi hills) Woody species (≥ 5 cm dbh)	3.26		0.061	Upadhaya *et al.*, 2008
Nongkrem (East Khasi hills) Woody species (≥ 5 cm dbh)	2.40		0.147	
Manipur Sacred Groves (4 numbers)				Khumbongmayum *et al.*, 2005
Tree species	1.79-3.17	1.89-2.25	2.77-3.13	
Shrubs	0.55-0.83	0.93-0.99	0.92-0.96	
Herbs	0.07-0.59	0.11-0.16	0.06-0.50	

The species diversity in the sacred groves found in different northeastern states varied widely. The Shannon index of species diversity for tree species varied from a low of 1.79 in a Manipur sacred grove to a high of 3.55 in a well protected sacred grove at Raliang in Meghalaya (Table 6.2). In general diversity of woody species was higher than the understorey components of the community in almost all the groves. As discussed above, it is difficult to ascertain the causes of variation in species diversity in different groves, but climatic condition, availability of favourable niches for species with varied regeneration guild and human disturbance played major role in influencing species diversity in the sacred grove community (Upadhaya *et al.*, 2003). In Manipur sacred groves biotic influences were chiefly responsible for determining species diversity and dominance distribution in the community (Khumbongmayum *et al.*, 2005). The forests in which trees of only one species constitutes 50-80 per cent of the canopy trees have often been considered as low-diversity forests (Connell and Lowman 1989). On the basis of this criterion the sacred groves may be regarded as high-diversity forest since much larger number of tree species was present in the canopy layer.

The relative importances of species in a community differ widely. The importance of a species has been conceived as dependent on the fraction of niche space the species occupies and the fraction of community resources it uses. The measure of importance is expressed as the species' fraction of density, biomass, productivity etc., in the community as a whole. The importance of species in a community arranged in progression from most to least important species results in to importance-value or dominance-distribution curve, which takes at least three forms. Higher the diversity greater is the equitability or evenness in distribution of dominance in the community. In most of the undisturbed sacred groves dominance-distribution curve exhibited log-normal distribution pattern showing high equitability or diversity and low dominance in the community (Figure 6.2).

6.4 Keystone Species

A species that have a disproportionately large effect on other species in a community is termed as keystone species. They play a central and critical role in maintenance of the community structure and it's functioning by virtue of their uniquely important traits and attributes. If an ecosystem can be returned to a state in which keystone species flourish then all other species, which depend on it, will also flourish. One of the simplest ways of managing any natural ecosystem is by identifying species, which play the key role of holding together the entire community or ecosystem. Loss of vital function and changes in the ecosystem structure would follow if they were removed from the system. These species are 'responsible' for the existence of an ecosystem of certain type and create possibilities for the existence of other types of communities.

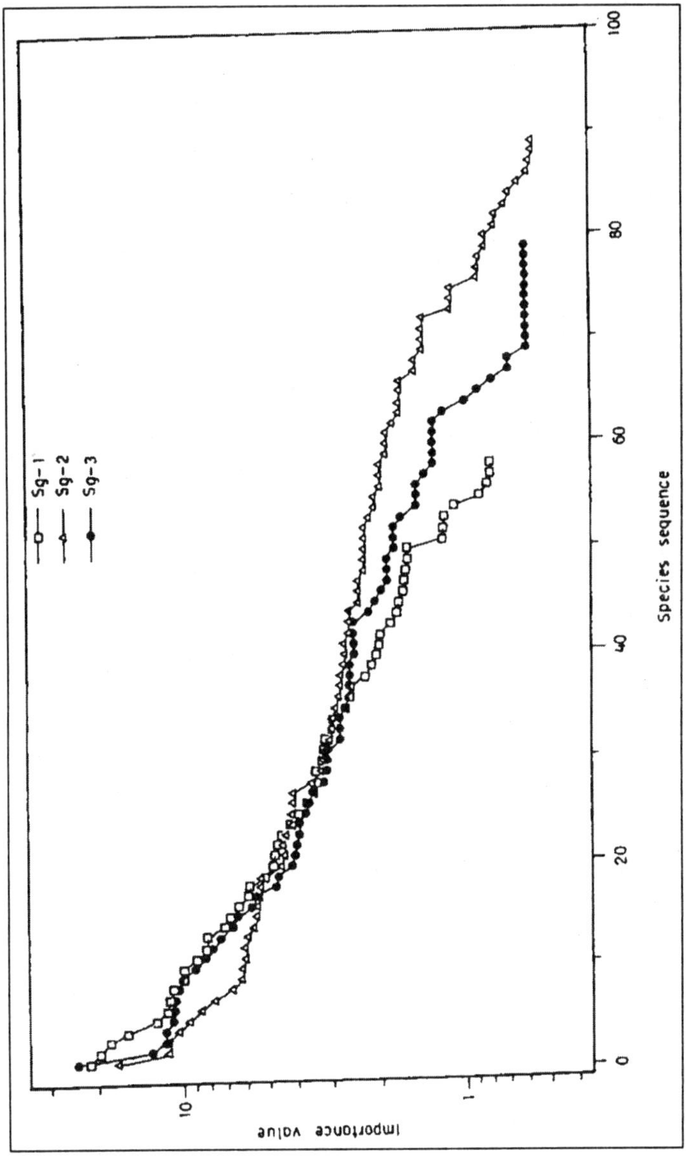

Figure 6.2: Dominance-Distribution Curves of Woody Species in Sacred Groves of Jaintia Hills (Sg-1 Khloo Paiu Ram Pyrthai, Sg-2 Urkhla, Sg-3 Khloo Langdoh) (after Jamir 2000)

Table 6.3: Some Keystone Species Identified in lalong Sacred Grove Based on Growth Form, Dominance, Litter Quality, and their Key Ecological Role in the Sacred Grove (after Law 2002)

Tree Species	*Growth Form*	*IVI Rank*	*Leaf Shedding*	*Litter Quality C : N*	*Key Ecological Role*
Engelhardtia spicata	Canopy tree	2	Seasonal	26.79	Pioneer/important for ecosystem reclamation and biodiversity rehabilitation/ supports epiphytes
Betula alnoides	Canopy tree	3	Seasonal	14.75	Pioneer/important for ecosystem reclamation and biodiversity rehabilitation/supports epiphytes
Castanopsis purpurella	Canopy tree	4	Continuous	27.01	Food resources for rodents/squirrels/important keystone resource/supports epiphytes
Pseudostreblus indica	Canopy tree	5	Continuous	25.37	Food resources for birds and wild animals/ important keystone resource/supports epiphytes
Castanopsis tribuloides	Canopy tree	8	Continuous	26.01	Food resources for rodents/squirrels/important keystone resource/supports epiphytes
Syzygium tetragonum	Canopy tree	9	Continuous	30.38	Food resources for birds and wild animals/ important keystone resource/supports epiphytes
Pithecellobium monodelphum	Subcanopy tree	11	Continuous	13.27	Supports Nitrogen fixing bacteria/important in nutrient cycling

Table 6.4: Some Keystone Species Identified in Raliang Sacred Grove Based on Growth Form, Dominance, Litter Quality, and their Key Ecological Role in the Sacred Grove (after Law 2002)

Tree Species	*Growth Form*	*IVI Rank*	*Leaf Shedding*	*Litter Quality C : N*	*Key Ecological Role*
Sarcosperma griffithii	Canopy tree	1	Continuous	37.01	Food resources for birds/wild animals/ important keystone resource/supports epiphytes
Actinodaphne obovata	Canopy tree	2	Continuous	23.05	Food resources for birds/insects/important keystone resource/supports epiphytes
Prunus jenkensii	Canopy tree	3	Continuous	22.70	Food resources for birds/wild animals/ important keystone resource/supports epiphytes
Neolitsea cassia	Canopy tree	4	Continuous	25.19	Food resources for birds/important keystone resource/supports epiphytes
Ficus virens	Canopy tree	5	Seasonal	26.06	Food resources for birds/rodents, squirrels/ important keystone resource/supports epiphytes
Ficus altissima	Canopy tree	7	Seasonal	30.47	Food resources for birds/rodents, squirrels/ important keystone resource/support epiphytes
Acer laevigatum	Canopy tree	8	Seasonal	20.03	Supports epiphytes/gap colonizers
Knema angustifolia	Canopy tree	10	Continuous	28.97	Food resources for birds/insects/important keystone resource/support epiphytes

The keystone species are usually noticed when they are removed or disappear from an ecosystem, resulting in dramatic changes to the rest of the community. Identifying keystone species is often difficult and attempts to develop a set of species traits that would determine keystone interactions have been thus far elusive. Keystone species emphasizes processes that directly (*e.g.* predation, competition) rather than indirectly (*e.g.* nutrient cycling) control biodiversity. They included organisms that (i) control potential dominants, (ii) provide critical resources, (iii) act as mutualists, and (iv) modify the environment.

Law (2002) has attempted to identify keystone species or keystone groups in two sacred groves of Jaintia hills (Table 6.3, 6.4), Meghalaya based on the following parameters:

1. Importance value index (IVI),
2. Position in the canopy,
3. Growth form/life form,
4. Association with other important species (epiphytic growth), 'mobile links', (*e.g.* pollinators, dispersers), capacity to supply food resources for faunal diversity,
5. Successional status/light-demanding or shade-tolerant species,
6. Biomass and nutrient allocation pattern to shoot and root,
7. Shoot/root ratio of seedlings, Leaf mass per unit area (LMA), and
8. Nutrient input to the system (litter quality, nutrient release and nitrogen fixation).

As discussed above the species listed in Table 6.3 and 6.4 do not fulfill the criteria laid down for identifying the keystone species *i.e.*, having direct effect on ecosystem processes that control biodiversity. Nonetheless, they provided critical food and space resources to a large number of animal and plant species and strongly influenced the forest microenvironment by changing light regime and resource quality of litter on the forest floor, which together has the potential to affect the tree regeneration and other vital processes such as litter decomposition and nutrient cycling in the ecosystem.

Chapter 7

Tree Regeneration

The process of replacement of old trees by young ones is called regeneration. Regeneration is necessary for growth and perpetuation of forest. Natural regeneration of trees takes place either through seeds or through vegetative parts (sprouting from stumps, rhizomes or roots) or by both. Successful regeneration through seeds depends upon seed production, seed dispersal, germination and seedling growth and survival. All these processes are affected by large number of internal factors specific to a species and external factors such as temperature, amount and distribution of rainfall, soil and light condition and microsite heterogeneity on the forest floor etc.

Seed production in trees varies from species to species and depends on age of the tree and site conditions. In many species only a small fraction of ovules develop in to ripe seeds due to pollination failure, deficiency of resources, and predation and failure in ovule development. Species differ from one another in seeding habit; some produce seed every year, some in alternate years and some do so after long intervals.

Before the seeds are dispersed from mature fruits, birds and frugivorous animals eat them and often help in their seed dispersal. Trees show different dispersal mechanism of seeds. These mechanisms increase the chance of seeds to reach to the favourable microsites for their successful germination and establishment. The important natural agencies involved in seed dispersal are wind, water and animals. Some seeds are dispersed by explosive mechanism of pods. In tropical rain forest tree seeds are dispersed primarily by animals.

Germination of dispersed seeds depends on their viability, dormancy behaviour and availability of favourable environmental condition on the forest floor. There are two broad categories of seeds with respect of their storage physiology, orthodox seeds, which retain viability for long periods if stored in sealed containers at –18°C or lower temperature and dried to moisture content of about 10 per cent and recalcitrant seeds which are killed if their moisture content is reduced below some relatively high value. Even if stored moist they have only short longevity. Majority of tree species have orthodox seeds and many primary rain-forest trees have recalcitrant seeds.

Seed dormancy staggers germination and helps in survival of seedlings in many species. However, seeds of many rain forest trees lack dormancy. This has serious implications for the natural regeneration. Sometimes dormancy is enforced because of unfavourable external factors. Among those, temperature and light are important. Micro-topography and favourable moisture condition are other important factors, which influence germination on the forest floor. Micro-topography of the forest floor determines the density of safe-site. The effect of micro-sites on germination is controlled by the amount of contact between seed and soil surface and on the tension with which water is held in soil. Therefore large seeds do not germinate unless they are buried beneath the litter or mineral soil.

Successful establishment of seedlings is an important prerequisite for regeneration. Large-scale mortality of seedlings is a common phenomenon in most tree species. Several factors are responsible for seedling mortality on the forest floor. Notable among them are drought, frost, high temperature and low light intensity, soil depth, nutrient status, soil reaction, soil moisture, soil aeration, pathological and entomological factors, grazing and trampling by wild animals, and competition and growth pattern and hardiness towards frost, drought and fire.

Natural regeneration of trees in the humid tropical forests is highly complex and least understood process. These forests are considered as the dynamic systems of ever changing mature, gap and building phases in which a particular combination of species that forms the dominants of a given small area is constant neither in space nor in time. The combination of species at a given place and time is succeeded, not by a similar combination but by a different one and no combination of species can be in permanent equilibrium with the environment.

Regeneration of trees from seeds in a closed canopy forest depends on the occurrence of gaps in the canopy. The conditions in the closed canopy forest are intensely inimical to the recruitment of new individual. Gaps provide habitats in which competition is reduced or absent and resources like light is available that may initiate seed germination and help establishment of seedlings.

Barik *et al.* (1992) studied creation of gaps and micro-environmental condition in canopy gaps and forest understorey in a sacred grove of Meghalaya representing humid subtropical broadleaved forest. They found that canopy gaps are created due to death of big trees or by snapping of bole and branches by the action of windstorm, rain, hail, lightning, land slip etc. Depending on the factors responsible for creation of gaps, the shape and size of the gaps vary widely. For example, trees felled by wind usually have their roots torn from the ground and forced in to a mound, with a corresponding depression. The size variation in the gap and its frequency are also important for regeneration of trees.

Microenvironment in gaps differs from the forest understorey. The light, temperature and moisture regimes in the gaps are radically different from the forest understorey. Sometimes there is marked local alteration in micro-relief and soil profile in the gaps. The gap microenvironment initiate a set of plant responses such as germination of dormant seeds in soil and upward growth of existing seedlings and saplings whose growth was suppressed by shade cast by over head canopy. Species also differ in their responses to gap opening in the forest depending on their light requirement; some are light demanding or pioneer species and others are shade bearing or tolerant or primary species. Each of these two groups of species has different germination and growth requirements.

Natural regeneration of trees also occurs by coppice and root suckers. Adventitious or dormant buds form new independent plants from the parent plant. The adventitious buds originate from the callus tissues around the wound while the dormant buds are stimulated to produce new shoots when physiological balance is upset by some injury. 'Stool shoots' generally arise from the short-lived adventitious buds formed between the wood and the bark of the stump. Large number of tree species has the ability to regenerate in this way.

7.1 Seed Production

Seed production of trees is influenced by various extrinsic factors such as resource availability, pollination agents, predation on flowers, fruits and leaves and by climatic conditions as well as by intrinsic factors such as age and size of the plant and its genetic constitution (Winn and Werner 1987). The patterns of seed production differ in the same species that are adapted to different microhabitats; and individual trees within a population also vary enormously in the size of their fruit crop (Howe 1982). Seed production among the tree population also varies due to differences in the number of seed produced per reproductive individual (Schupp 1990).

Seed production in dominant tree species, *Lithocarpus dealbatus*, *Quercus griffithi* and *Schima khasiana* in the Mawphlang sacred grove near Shillong significantly increased with increase in age (girth) of fruiting trees and varied from year to year (Barik *et al.*, 1996a) (Table 7.1). Similar year-wise variation in seed production has also been reported by Grubb (1977), Clark and Clark (1987) and many workers as cited by Schupp (1990) in both shade tolerant and light demanding pioneer species. The ultimate cause of cycle of alternate bearing is not known, but most hypotheses suggest that resource limitation in trees as its main cause (Monselise and Goldschmidt 1982, Fenner 1991). Howe (1982) and Janzen (1971) have argued that year-wise variation in heavy fruiting events ensures the recruitment of a particular species despite heavy seed and seedling predation by herbivores. Heavy fruiting may also occur because of prolonged depression of minimum night temperature, increasing length of dry season and a dry sunny year following a wet year.

7.2 Periodicity of Tree Seed Fall

The importance of seed fall has been increasingly recognized as an essential process that determines structure and dynamics of forest community (Hamilton 1999; Harms *et al.*, 2000). For any site, the seed community reflects the input of seeds of local origin, as well as movement into an area by long distance immigrants (Harper 1977; Martinez-Ramos and Soto-Castro 1993). At the community level expected seed input are difficult to assess accurately because individuals and species seed fall patterns are highly variable both in space and time. Adding to the complexity, reproduction often does not occur annually. It is highly variable in different years, and in populations of forest tree species (Greene and Johnson 2004).

Table 7.1: Year-wise Variation in Seed Production (Mean number x 10^{-3}) by Trees of Different Dbh Classes in Mawphlang Sacred Grove

Species	*Collection Period*	*Dbh Class (cm)*		
		10-20	*20-30*	*30-40*
Lithocarpus	November 1988	0.3±0.1	0.8±0.3	2.0±0.9
dealbatus	November 1989	0.6±0.3	1.9±0.3	3.3±0.6
	November 1990	0.7±0.3	2.4±0.9	4.0±1.6
Quercus	December 1988	9.1±2.8	19.9±8.2	43.0±8.2
griffithii	December 1989	3.5±1.6	9.5±3.8	22.7±10.4
	December 1990	7.4±4.1	14.3±6.3	37.7±12.3
Schima	January 1989	6.6±1.7	10.7±1.7	17.5±6.3
khasiana	January 1990	6.4±3.4	14.8±4.2	20.5±7.6
	February 1991	5.3±2.2	8.4±4.2	13.2±8.7

± S.D.

Rain forest trees, particularly canopy trees, those belonging to either A or B strata have times of maximum and minimum flowering and fruiting leading to seasonal seed production. They may flower either annually, biennially or even thrice a year, though seeds are not always set, or if set, they produce good seeds in fair quantity at least once in three years. Sometimes no fruits are set after flowering because of several factors like unfavourable weather condition for fertilization and damage by birds or other animals which feed on unripe fruits. The peaks and troughs of seed setting and seed fall are not always clearly defined, but reoccur at about the same time every year (Richards 1996).

Seed fall pattern of tree species in Swer and Shorarim sacred groves representing humid subtropical broadleaved forest in Meghalaya was studied by Nongbri (2007). In both the sacred groves the duration of seed fall in a given species ranged from 1-5 months but at community level seed fall occurred throughout the year (Fig 7.1 and 7.2). Spring (March-May) was the peak period of seed fall when majority of species produced seeds (Figure 7.3). Seeds produced during this season had better chance of germination in the ensuing rainy season while those formed during autumn and winter had to undergo a relatively longer period of dormancy until the condition becomes favourble during rainy season. In *Castanopsis purpurella* and *Acer laevigatum* fruiting occurred between October–November (autumn) and in *Engelhardtia spicata* and *Betula alnoides* seeds were formed during by March-April (spring). Their seeds remain dormant until the next rainy season. However, *Sarcosperma griffithii, Syzygium tetragonum* and *Neolitsea cassia* which produced seeds in July-August, germination occurred within few days of seed fall.

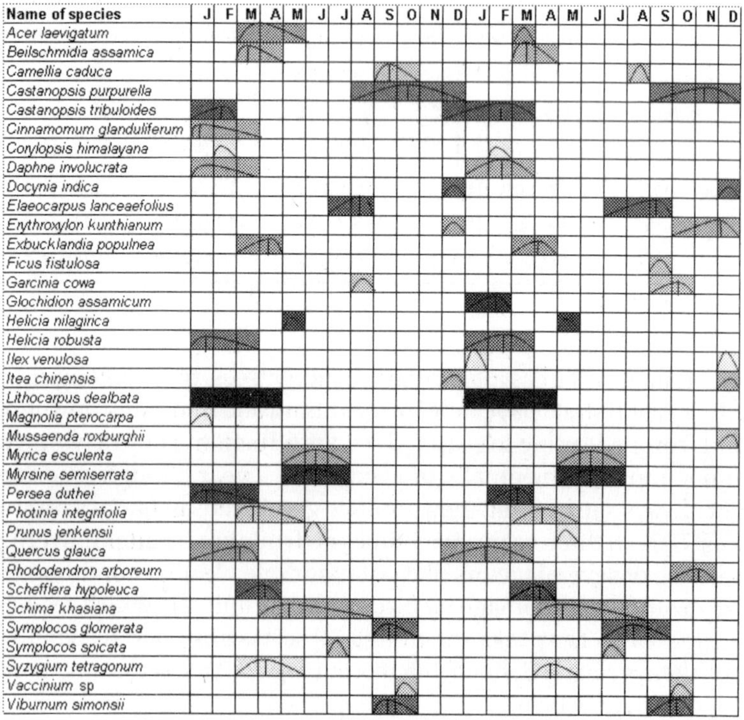

Figure 7.1: Periodicity of Tree Seed Fall in Sohrarim Sacred Grove During 2003-2004

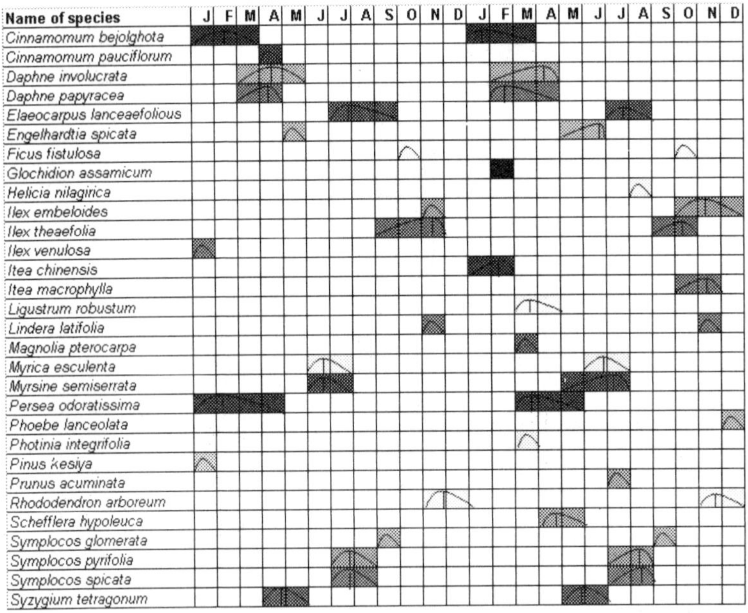

Figure 7.2: Periodicity of Tree Seed Fall in Swer Sacred Grove During 2003-2004

At community level seed fall showed seasonal as well as yearly variation in both the groves. In both the forests majority of the species (18 species in Sohrarim and 12 species in Swer) shed their seeds during spring season (March–May). Seed fall in few species occurred either during winter season in Sohrarim or during rainy and autumn seasons in Swer sacred grove (Figure 7.3).

7.3 Seed Dispersal and Fate of Seeds

Both in Shorarim and Swer groves majority of tree species produced animal dispersed seeds. In Sohrarim, only 5 canopy tree species *viz., Rhododendron arboreum, Exbucklandia populnea, Acer laevigatum, Mussaenda roxburghii* and *Schima khasiana* had wind disseminated seeds. In Swer forest three wind disseminated seed species *viz., Rhododendron arboreum, Pinus kesiya* and *Engelhardtia spicata* were recorded. It was observed that generally the canopy species produced wind-disseminated seeds, which were shed during dry winter season (December to February) or before pre-monsoon showers during April and May.

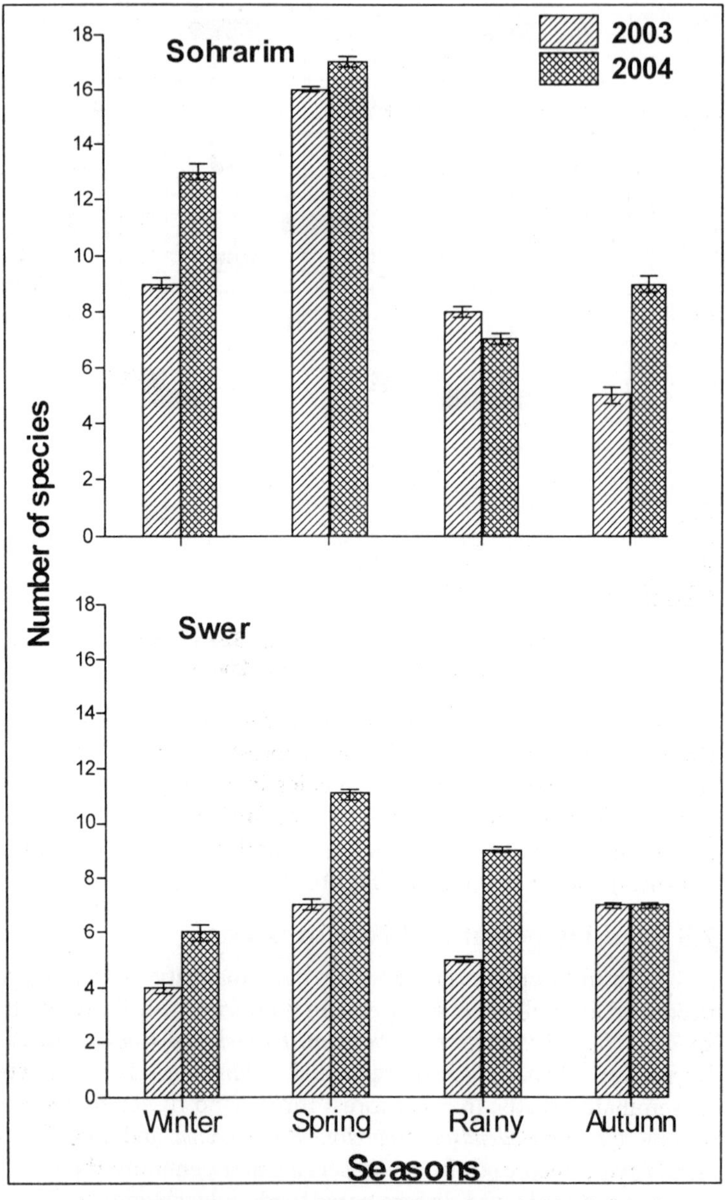

Figure 7.3: Seasonal Seed Fall of Tree Species in Sohrarim and Swer Sacred Groves (Winter = December–February; Spring = March–May; Rainy = June–August; Autumn = September–November)

In Mawphlong sacred grove number of seeds per unit area of the forest floor decreased significantly with the increase in distance from the parent tree crown. Such a seed dispersal pattern was observed in *Lithocarpus dealbatus* and *Quercus griffithii*. Their highest number of seeds was recorded within 5 m radius of the tree. During the seed fall period, 52 per cent–60 per cent of the total seeds produced by *Lithocarpus dealbatus* disappeared. The corresponding value for *Quercus griffithii* ranged between 84 per cent and 91 per cent. Even during the post-seed fall period, more than 95 per cent of the remaining seeds disappeared. In general, more than 98 per cent of the total seeds produced by the two species of oak disappeared from the forest floor; this included the seeds damaged by rodents and insects. It was estimated that 0.1-0.5 per cent seed got rotten and less than 1 per cent (0.1-0.8 per cent) could have germinated (Barik *et al.*, 1996a). Seed predation decreased with the increase in distance from the tree. Rapid disappearance of seed from the forest floor has been ascribed to transportation by mice, voles and other rodents (Jensen 1985, Kikizawa 1988) which consume and/or scatter and hoard seeds beneath the deep litter layer or below ground. The acorns of *Lithocarpus dealbatus* and *Quercus griffithii* are highly prone to herbivory by rodents in the forest owing to their high food content and absence of any strong defense against herbivory except a thickened seed coat and rounded form.

Small-scale heterogeneity in biotic and abiotic variables of environment divides forest floor into a mosaic of microsites, some of which favour germination of a particular group of species depending on their germination requirements. For instance germination *Lithocarpus dealbatus*, *Quercus griffithii* and *Schima khasiana* on moss layer was much higher than in the other microsites on the forest floor (Barik *et al.*, 1996a, Tripathi and Khan 1990). The lowest germination of these species was observed on the litter layer (Table 7.2). The moss layer which retains considerable amount of moisture provided most favourable microsite for germination while litter which quickly dries up proved to be unfavourable for those seeds which fall on it. The germination was poor during winter season when soil moisture content and ambient temperature remain low. Several workers studying the seed ecology of *Quercus* species (Barik *et al.*, 1996a, Tripathi and Khan 1990) have reported that the unconsumed scatter-hoarded fraction of the disappeared seed population often germinates, owing to favourable moist condition below the litter, and give rise to a sizeable population of seedlings.

Table 7.2: Effect of Microsites (Moss layer-ML; Over litter layer-OL; below the litter layer-BL, cleared forest floor-CF) and Seed Mass on Percentage Germination (mean±SE) of Few Dominant Tree Species of the Sacred Groves

Species	*Seed Weight Class (mg)*	*Microsites*			
		ML	*OL*	*BL*	*CF*
Lithocarpus	300-500	20±2.9	5±1.7	10±4.4	10±1.7
dealbatus	500-700	25±1.7	5±2.9	10±1.7	10±1.7
	700-900	30±4.4	10±1.7	15±2.4	15±1.7
	900-1100	30±2.9	10±1.7	15±2.4	20±4.4
Quercus	300-500	35±1.7	10±1.7	20±4.4	15±2.4
griffithii	500-700	40±2.4	15±1.7	25±2.9	20±2.9
	700-900	50±2.9	20±12.4	30±2.9	25±1.7
	900-1100	60±2.9	20±1.7	30±2.4	25±1.7
Schima khasiana		50±3.5	5±2.2	25±2.2	35±1.6

7.4 Tree Seedling Growth

Study of growth and nutrient allocation in seedlings of seven tree species found in two sacred groves of Jaintia hills in Meghalaya revealed that some species (*Pithecellobium monadelphum, Camellia caudata, Microtropis discolor)* had higher root growth, whereas others (*Engelhardtia spicata, Betula alnoides)* had higher shoot growth (Table 7.3). The important canopy species such as *Knema angustifolia* and *Pseudostreblus indica* showed a similar increment in both root and stem.

Resource allocation to the below ground and above ground parts strongly influences fitness of the species. The difference in shoot/root biomass ratio among forest tree species is an important character in determining the adaptive strategies of species along successional gradient. Leaf biomass per unit area is usually higher in the canopy tree species than in the subcanopy and undercanopy species. The success of a species in an environment can be predicted by looking at its relative apportionment of biomass or nutrients in the above and below ground parts. Some plant species show high proportion of their biomass allocation to the root system allowing them to acquire more soil resources than plants with less root biomass and thus are superior competitors for soil nutrient resources. However, by allocating more biomass to the roots the plant species necessarily have less allocation to stem, leaf and reproductive buds and are inferior in their ability to utilize light and to disperse seeds.

On account of greater allocation of biomass to the shoot system *Betula alnoides, Castanopsis purpurella, Engelhardtia spicata, Syzygium tetragonum, Actinodaphne obovata, Neolitsea cassia, Sarcosperma griffithii* and *Prunus jenkensii,* all canopy tree species may be considered shade tolerant and therefore better adapted to forest understorey. On the contrary allocation of biomass to shoot and root was almost equal in *Camellia caudata, Microtropis discolor, Pithecellobium monadelphum* and *Pseudostreblus indica* suggesting that these could grow well both under shade and light condition (Table 7.3).

7.5 Effect of Seed Weight on Germination and Growth and Survival of Seedlings

The wide differences in seed mass among species are regarded as an important aspect of their reproductive strategy (Grubb 1996). Seed mass plays an important role in the establishment of seedlings principally under conditions where resources are scare (Grubb and Coomes 1997, Grubb and Burslem 1998). Foster (1986) listed several potential advantages of large seeds of tropical trees from moist forests and argued that large seed reserves might be used for the construction of large amount of photosynthetic tissue to maintain a positive net energy balance, or might also allow quick seedling growth for reaching higher light intensity strata. The differences in seed mass have also been related to the ecological conditions in which plants establish, with species from open habitats having lower seed mass than species from more closed habitats (Foster and Janson 1985). Large seeded species have an advantage in competitive environments (Gross 1984) when seedlings experience defoliation (Armstrong and Westaby 1993) or moisture stress (Baker 1972).

Food reserve in seeds increases with seed weight. Conflicting reports are available on the effect of seed weight on germination

Table 7.3: Growth (Dry weight (mg)/plant) of Two-Year Old Tree Seedlings in Ialong and Raliang Sacred Grove

Tree Species	*Root*	*Stem*	*Leaf*	*Total*	*S/R ratio*
Ialong sacred grove					
Betula alnoides	560±2.1	1125±3.2	467±4.2	2152	2.84
Camellia caudata	1299±7.0	1021±11.1	397±5.0	2717	1.09
Castanopsis purpurella	837±10.1	1215±10.5	678±4.4	2730	2.26
Engelhardtia spicata	715±6.1	1275±6.4	971±5.0	2961	3.14
Microtropis discolor	718±2.6	480±3.3	257±3.1	1455	1.03
Pithecellobium monadelphum	1075±8.5	690±11.0	587±5.7	2352	1.19
Pseudostreblus indica	1070±5.3	750±10.1	597±3.3	2417	1.26
Syzygium tetragonum	760±11.2	1310±13.1	715±1.1	2785	2.66
Raliang sacred grove					
Acer laevigatum	757±0.6	767±9.1	495±4.5	2019	1.55
Actinodaphne obovata	1157±13.1	1469±19.0	1031±1.1	3657	2.16
Knema angustifolia	957±31.2	1011±8.5	540±9.1	2508	1.62
Neolitsea cassia	780±14.0	1345±12.1	360±5.0	2485	2.19
Prunus jenkensii	775±15.2	1233±16.2	497±6.3	2505	2.23
Sarcosperma griffithii	460±10.0	775±8.3	459±8.4	1694	2.68

behaviour. For instance, large seeds may germinate to higher percentage than those of small seeds (Tripathi and Khan 1990, Bhuyan *et al.*, 2000, Khan and Uma Shankar 2001), and at times small seeds may also germinate to higher percentage than large seeds (Marshell 1986) or germination may be independent of seed size (Gross and Kromer 1986, Perez-Garcia *et al.*, 1995). Within a species, heavier seeds may take lesser time for germination than lighter seeds (Barik *et al.*, 1996a), lighter seeds may germinate earlier than the heavier seeds (Murli 1997, Khan *et al.*, 1999) or germination time may not be dependent on seed weight (Perez-Garcia *et al.*, 1995). Germination experiment on *Prunus jenkensii* in the laboratory showed that though heavier seeds took longer time to germinate but their percentage germination was significantly higher than the lighter seeds. The seedlings coming out of the heavier seeds had greater seedling vigour than those coming out from lighter seeds (Upadhaya *et al.*, 2007) (Table 7.4).

Table 7.4: Germination Behaviour of Seeds and Seedling Survival and Growth After One Year of Germination of *P. jenkinsii* (After Upadhaya *et al.*, 2007)

Seed Weight Class (g)	*Germination (Per cent)*	*Number of Days for Initiation of Germination*	*Seedling Survival (Per cent)*	*Leaf Area (cm^2)*	*Dry Weight/ Plant (g)*
Light (<1.5)	28	39	42	44.4±3.2	0.291±0.012
Intermediate (>1.5-2)	32	49	63	65.4±4.5	0.379±0.022
Heavy (>2)	45	75	80	73.3±2.3	0.433±0.013

Heavier seeds confer an advantage to seedlings for their better survival and growth due to large reserves of nutritive substances. Tripathi and Khan (1990), Barik *et al.* (1996a) argued that seedlings emerging from seeds of varied mass exhibit differential competitive performance. Seedlings arising out of large and heavy seeds show better emergence, survival and growth than the small seeds (Bonfil 1998), while light seeds confer the advantage of greater dispersal efficiencies (Ganeshaiah and Uma Shaanker 1991, Hedge *et al.*, 1991).

7.6 Effect of Gaps and Understorey on Germination and Seedling Survival

In closed canopied forests like those represented by sacred groves at higher elevation in Meghalaya, tree fall gaps play an important role in successful regeneration of tree species by influencing seed germination and survival and mortality of their seedlings. The behaviour of tree seedlings in tree fall gaps and understorey also depends on species' niche specialization (Parrish and Bazzaz 1982) and difference in their ability to utilize the available resources in the gaps (Denslow 1980). Accordingly, species have different regeneration strategies to colonize different microsites in the forest understorey and gaps. The small gaps and understorey are occupied primarily by shade-tolerant, primary or mature phase species (Richards 1996) while the large gaps are colonized by light-demanding, pioneer or shade-intolerant species (Brokaw 1985).

Study of microenvironment in tree fall gaps and forest understorey of Mawphlang sacred grove in Meghalaya revealed that photosynthetically active radiation markedly increases from the understorey to the large gaps. Both gaps and understorey received maximum radiation during the dry winter months when several tree species shed their leaves and minimum during the wet summer season when almost all tree species are in full foliage. Though soil moisture did not vary between gaps and understorey, litter accumulation declined from the understorey to the large gaps and recorded highest values during winter and lowest during the wet summer.

The seedlings of oak species (*Lithocarpus dealbatus, Quercus glauca* and *Quercus griffithii*) are recruited mainly in the small gaps and forest understorey, while those of *Schima khasiana* are abundant in the larger gaps (Table 7.5) and in the open disturbed sites (Khan *et al.*, 1986, Barik *et al.*, 1992). The better germination of *S. khasiana* seeds in tree fall gaps than in the adjacent understorey, and recruitment of a large number of its seedlings in the large gaps indicates that better light environment in the large gaps favours its recruitment. Based on the seed germination and survival pattern and growth response of seedlings of three oak species and *Schima khasiana* in gaps and understorey, Rao *et al.* (1997) concluded that the seedlings were adapted to different light environments depending on their optimum requirements (Table 7.5).

Table 7.5: Mean Population Density (±SE) and Per cent Mortality of Seedlings of some Tree Species in Mawphlang Sacred Grove (After Rao *et al.*, 1997)

Species	*Canopy Openings*	*Seedling Density (Plants/ha)*		*Seedling Mortality (%)*
		August 1988	*August 1990*	*August 1990*
Lithocarpus dealbatus	Forest under storey	725±95	322±64	55.6
	Small gaps	526±84	195±36	62.9
	Large gaps	143±14	23±10	83.9
Quercus glauca	Forest under storey	391±50	156±49	60.1
	Small gaps	124±33	32±19	74.1
	Large gaps	64±32	12±6	81.3
Quercus grifithii	Forest under storey	250±40	142±44	56.8
	Small gaps	139±56	46±17	66.7
	Large gaps	68±27	18±8	73.5
Schima khasiana	Forest under storey	125±33	0	100.0
	Small gaps	192±25	24±13	87.5
	Large gaps	248±26	188±42	24.2

All three species of oak suffered greater mortality in the gaps than the understorey. *Schima khasiana* had significantly higher mortality in small gaps and understorey than the large gaps. Thus regeneration of the three species of oak was better in the understorey and small gaps, while that of *Schima khasiana* and *Engelhardtia spicata* were favoured by canopy openings. This response of tree seedlings to the conditions in the understorey and gaps indicates that different tree species are adapted to different light regimes depending upon their optimum requirements (Table 7.6). This appears to be an effective mechanism for promoting species coexistence and diversity in the community.

Table 7.6: Favoured Regeneration Niche of Tree Species in Sacred Groves as Indicated by their Seedling Response Attributes (Rao *et al.*, 1997, Law 2002)

Species	*Germination*	*Survivorship*	*Relative Shoot Elongation Rate*	*Leaf Area*
Small gap and under storey species				
Lithocarpus dealbatus	Gaps	Understorey	Understorey	No preference
Quercus glauca	No preference	Understorey	Small gaps/under storey	Large gaps
Quercus griffithii	No preference	Understorey	Understorey	Understorey/small gaps
Large gap species				
Schima khasiana	Gaps	Gaps	Large gaps	Large gaps
Engelhardtia spicata	Gaps	Gaps	Large gaps	Large gaps

Based on seedling survival and population structure of tree species in the two sacred forests of East Khasi Hills, Nongbri (2007) categorized them into three groups *i.e.* (i) 'shade tolerant'- those species whose seedlings could survive in the forest understorey and had a reverse J-shaped population structure, (ii) 'shade-intolerant'– those species whose seedling could not survive in the forest understorey and were mostly present only in adult stage and (iii) 'intermediate type'- those species whose seedlings were able to survive under the canopy, and had sporadic distribution of individuals in the populations. In both the groves there was abundance of shade intolerant species (23-28 species), followed by shade tolerant species (14-19 species) and intermediate type species (6-8 species).

Tree seedling recruitment on the forest floor was a seasonal phenomenon, starting with the onset of pre-monsoonal showers in April and culminating with the cessation of rain and lowering of atmospheric temperature from October onwards. There was no recruitment in most of the studied species between November and February when the ambient temperature was low and surface layer of soil was dry (Rao *et al.*, 1997, Law 2002).

Based on overall population behaviour and seedling performance in undisturbed and disturbed stands of two sacred groves in Jaintia hills, species like *Actinodaphnae obvata, Prunus jenkensii, Castanopsis purpurella* and *Acer laevigatum* were regarded as light–flexible tree species since they exhibited luxuriant growth in the disturbed stand than in the undisturbed stand. On the contrary *Knema angustifolia* and *Pseudostreblus indica* which performed better in closed canopied condition were considered as shade-tolerant species. Seedlings of *Engelhardtia spicata* showed 100 per cent mortality and those of *Betula alnoides* did not show any sign of regeneration under the tree canopy and performed better in the disturbed areas. Therefore, these species behaved like pioneer species and played an important role in regrowth of sacred groves after disturbance (Law 2002).

7.7 Survival and Growth of Transplanted Seedlings in Gaps and Understorey

Odyuo (2004) investigated survival and growth of seedlings of *Podocarpus neriifolia, Acer laevigalum* and *Syzygium tetragonum* in gap and understorey and found that seedling survival and growth of all three species was poor in the understorey than in the gap. The seedlings in gap had higher dried mass than those in the understorey. In all three species root: shoot ratio was higher in the gap than in the understorey. The RGR of all three species was higher in gap than in the understorey (Table 7.7).

Table 7.7: Average Mortality (Per cent), Mean Dry Mass (g plant^{-1}) Accumulation, RGR (mg day^{-1} plant^{-1}) and Root Shoot Ratio of Tree Seedlings in Gap and Understorey After Two Years of Transplantation (After Odyuo 2004)

Species	*Light Conditions*	*Mortality*	*Dry Mass*	*RGR*	*Root Shoot Ratio*
Podocarpus	Gap	16	952	14	0.52
neriifolia	Understorey	37	160	12	0.33
Acer	Gap	58	516	11	0.70
laevigatum	Understorey	67	124	8	0.50
Syzygium	Gap	50	2838	9	0.35
tetragonum	Understorey	58	97	7	0.27

Regeneration of tree species in the sacred groves is therefore a very complex process. It is dependent on several factors starting from seed production, timing of its shedding, dispersal pattern, dormancy, predation and other losses and availability of favourable microsites for germination and seedling establishment. Among the environmental factors light, temperature and soil moisture played most important role in seed germination and seedling growth on the forest floor. All three were influenced by seasonal cycle and gap phase dynamics. Besides, the tree species in sacred forest had different requirement for light. Depending on which they occupied different spatial niches on the forest floor staring from deep shade in the understorey to more lighted large gaps and forest periphery.

It was observed that a very small percentage of seed produced by trees was able to germinate and produce seedlings. These seedlings suffered heavy mortality in the first year of their recruitment. The mortality percentage of the surviving seedlings decreased in the subsequent years. The seedling growth was dependent on the availability of light, water and nutrients for which they have to compete with the overhead tree species. As a result their growth remains arrested or they grow at a slow pace. This is manifested in the population structure of woody species in the sacred groves, which show preponderance of individuals of smaller girth.

Survival, growth and establishment of tree seedlings under different environment are mainly determined by the availability of light, water and nutrient resources and by the competitive ability or adaptive strategies of the species along these environmental gradients (Law 2002). Adaptability of plants to various environmental conditions is determined by their differences in morphological and physiological traits and attributes. Morphological attributes are predominantly associated with differences in capacity of dehydration avoidance, with root and leaf adaptations playing an important role in promoting better plant water balance. Development of vigorous root system is critical to the successful establishment of seedlings because soil water resources must be exploited to make up for the transpiration losses of rapidly growing seedlings (Harper 1977). The pattern of allocation of dry matter to root and shoot affects water relations; greater drought tolerance has been associated with smaller investments in shoot growth as compared to the root growth in numerous species (Abrams 1990).

Light is a major environmental factor that limits growth and survival of many forest species (Whitmore 1996). Gap formation due to tree fall or dropping of branches in the forest provides suitable micro-sites for recruitment of seedlings of pioneer or shade intolerant tree species. Gap increases environmental heterogeneity of forests in many ways. Light and soil parameters (moisture, nutrients) vary with gap size; the centers of large gaps are different from the centers of small gaps, which are different from the surrounding understorey (Runkle 1981). The entire process of seedling establishment starting from germination, seedling recruitment and their growth and development is strongly influenced by microsite heterogeneity and microclimatic conditions prevailing in the forest understorey and in the gaps (Barik *et al.*, 1992, Rao *et al.*, 1997).

In forest ecosystems, low light intensity often proves to be a limiting factor for seedling growth and plays a significant role in determining the outcome of the competitive struggle between seedlings and established plants. Neighbouring plants may interfere with and reduce each other's supply of light by direct interception

(Harper 1977). Light availability decreases during succession from early to late successional stage, as biomass increases (Parrish and Bazzaz 1982). Obviously, the species, which appear in later stage of succession, should have better ability to utilize light than those appearing in the earlier stages. Seedlings of primary or shade tolerant species perform better in the under-canopy or shade (Whitmore 1975). In a study from Borro Colorado Island, Welden *et al.* (1991) reported that growth of seedlings and saplings of majority of species was more rapid in areas of low canopy gaps where light was more than in the areas of high canopy trees. The emergent and dominant canopy layer species showed high sapling mortality rate than the less common species (Okuda *et al.*, 1997).

Field studies have left no doubt that desiccation is a major source of seedling mortality in the understorey of a wide range of forest types including rain forests. Water and nutrient deficits are often caused by slower rate of their supply to plants than the rate required for their maximum growth. Low nutrient availability and effects of neighbouring vegetation are other factors, which determine seedling growth and establishment (Weih and Karlsson 1999). Competition between roots of different species for uptake of water and nutrients depends on whether they coexist in the same soil zone or whether their demands are segregated spatially or temporally. Differences in the morphological and anatomical features of the root system are also important in this respect.

Chapter 8

Nutrient Dynamics: Role of Aboveground Litter, Fine Roots and Microbial Biomass

The structure and function of forest ecosystem is determined by the plant component more than any other living component of the system. The importance of plant control at the ecosystem level lies in their ability to transform kinetic solar energy into potential chemical energy and driving nutrients from soil back to the plants. In humid tropical forest ecosystem a high proportion of total stock of nutrients is held in shoot system due to its enormous amount of biomass and "leak-proof" efficiency of the forest floor and soil compartments that enables rapid recyclying of nutrients to the shoot system (Richards 1996).

The soil nutrients within the forest ecosystems flow through (*i*) absorption by roots and assimilation, (*ii*) allocation to biomass construction and maintenance, (*iii*) translocation from senescent tissues to metabolically active tissues of plants, (*iv*) return to soil through above and below ground litter and (*v*) microbial mediated release of inorganic nutrients into soils solution *i.e.* mineralization during organic matter decomposition (Barnes *et al.*, 1998).

In forest ecosystem nutrients entering from different sources such as mineral weathering, atmospheric deposition and biological fixation etc. may enter soil solution where they are absorbed by plant roots. Within the plant, absorbed nutrients participate in biochemical processes, and in some cases, these are translocated to other physiologically active parts prior to their shedding as litter. Most of the nutrients leaving the aboveground biomass as litter or as leachates reach the forest floor where they are decomposed by soil microorganisms. During the process of decomposition, soil bacteria, actinomycetes and fungi assimilate the organic compounds present in plant litter into their cells for biosynthesis and release inorganic nutrients into the soil. The decomposition of litter by microbes into inorganic ions and turnover of labile soil organic matter is affected by several factors such as litter quality, timing of litter input, and climate and soil conditions.

In humid tropical forest ecosystem a dense network of fine roots present on the forest floor and in the underlying mineral soil layer quickly absorb the soluble nutrients released during decomposition process and check nutrient loss from the system.

The soil microbes has relatively small standing stock of nutrients compared to soil organic matter and above and below ground litter, but it is a labile source of nutrients for plants, a pathway for incorporation of organic matter into the soil, and a temporary sink for nutrients. The amount of soil microbial biomass differ from one ecosystem to another depending on quality and quantity of leaf litter input (Bauhas *et al.*, 1998), rates of decomposition of organic matter (Melillo *et al.*, 1982), mineralization (Vitousek *et al.*, 1982, Zak and Pregitzer 1990) and nitrification (Finzi *et al.*, 1998, Lovette and Reuth 1999).

Some of the growth limiting nutrients like N and P in particular are lost from the soil through surface runoff and leaching when precipitation exceeds the evaporative loss from the soil surface. A small amount of N is also lost to the atmosphere through denitrification, a microbial-mediated process of reduction of nitrate to nitrous oxide or nitrogen (Tiedje 1988). These losses of limiting nutrients have the potential to influence the productivity of terrestrial ecosystems.

Studies pertaining to nutrients dynamics in sacred groves are highly fragmentary and limited primarily to the soil system, since uptake by plants and their accumulation in the vegetation is difficult to estimate without harvesting the vegetation, which in case of protected sacred grove is not possible. In Meghalaya, Kheiwtem and Ramakrishnan (1993) carried out studies on nutrient cycling in a sacred grove at Cherrapunji and recently Kamei (2007) investigated the nutrient dynamics in Swer sacred grove located about 25 km from Shillong on the way to Cherrapunji. In both the studies role of litter, fine roots and soil microbes in nutrient dynamics have been emphasized. Upadhaya *et al.* (2005) studied fine and coarse root dynamics in two sacred groves of Jaintia hills. The account given in the following pages is based on the results obtained in these studies.

8.1 Soil Nutrient Status

The soil in the Swer grove had sandy loam texture and acidic in reaction; the acidity being more pronounced during the rainy season (mean pH 4.06). Being located in high rainfall zone, soil in most sacred groves of Meghalaya is highly leached and is poor in nutrients than at other locations in north-eastern India (Table 8.1).

The nutrient concentration in the soil showed a distinct seasonal pattern as was evident from the analysis of NH_4-N, NO_3-N and PO_4-P. NH_4-N was the dominant form of inorganic nitrogen present in the soil and it varied widely from a low of 5 kg ha^{-1} during the wet rainy season to a high of 12 kg ha^{-1} during dry winter season. The NO_3-N concentration was lower than the NH_4-N. The seasonal trend was, however, similar in both NO_3-N and NH_4-N. The amount of PO_4-P was much lower than the total inorganic nitrogen. However, its seasonal trend was similar to the total inorganic nitrogen (Figure 8.1). All three nutrients exhibited a marked seasonality with a steady rise from September in the post-rainy month to the winter season (December- March) thereafter a gradual declined was noticed as the plants resumed growth during spring after the winter season. Their minimal amount was recorded during rainy season (June-July) when the possibility of losses from the soil was high and uptake by vegetation was at the peak.

8.2 Litter and Fine Root Production

The litter fall pattern in the Swer sacred grove is shown in Figure 8.2. Periodicity in litter fall largely follows seasonal cycle. It virtually starts in the dry winter month of December and steadily rises till the spring season in the month of March. A small peak observed in the month of August is an unusual feature attributed to heavy down pour recored during this month. Annual litter production in Swer

Table 8.1: Physico-chemical Properties of Soil of the Swer Sacred Grove

Soil Parameters	
Clay (per cent)	15.59±0.91
Silt (per cent)	21.64±0.44
Sand (per cent)	62.78±1025
Textural Class	Sandy Loam
Bulk Density (g cm^{-3})	0.85±0.01
Porosity (per cent)	67.9±0.15
Water holding Capacity (per cent)	61.09±0.42
Soil Moisture Content (per cent)	42.4±1.97
pH	4.54±0.06
Soil Organic Carbin (per cent)	2.88±0.09
Soil Organic matter (per cent)	4.97±0.15
Total Kjeldhal Niotrogen (per cent)	0.77±0.06
Available Phosphorus (µg g^{-1})	3.91±0.53
C/N Ratio	3.91±0.53

±S.D.

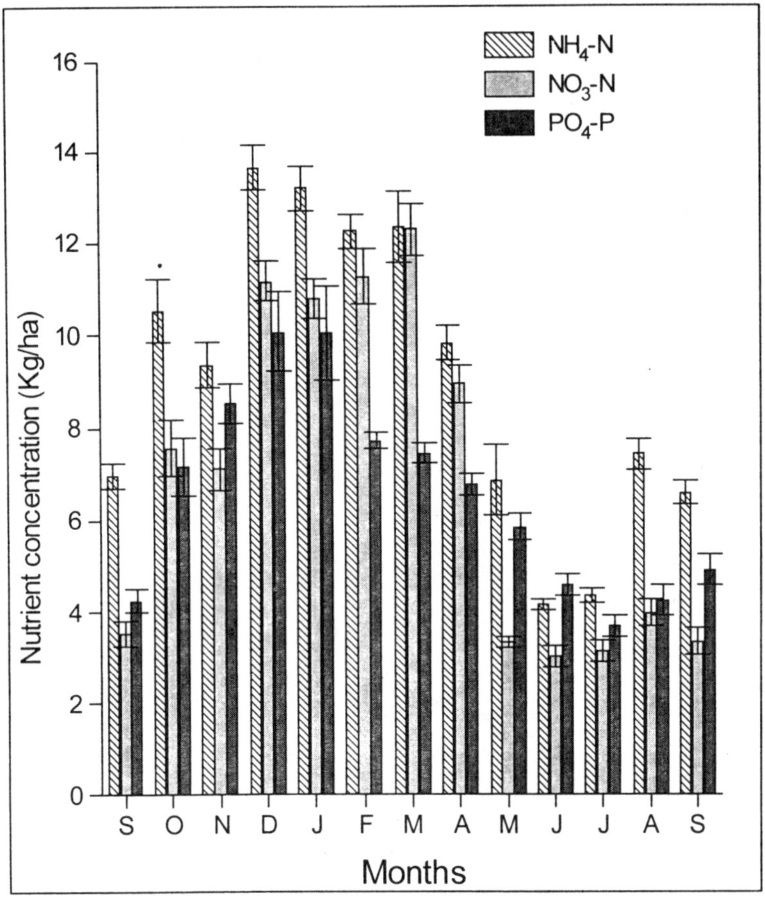

Figure 8.1: Monthly Variation in the Concentration of Soil Nutrients in Swer Sacred Grove

and Cherrapunji sacred groves ranged between 11371 and 11784 kg ha^{-1} (Khiewtam and Ramakrishnan 1993, Kamei 2007) with mean accumulation of about 5201 kg ha^{-1} on the forest floor. The proportion of leaf litter was about 75 per cent of the total litter fall as well as of the mean litter mass present on the forest floor.

Unlike litter, seasonal pattern of fine roots production was not very distinct (Figure 8.3). However, it gradually increased from spring

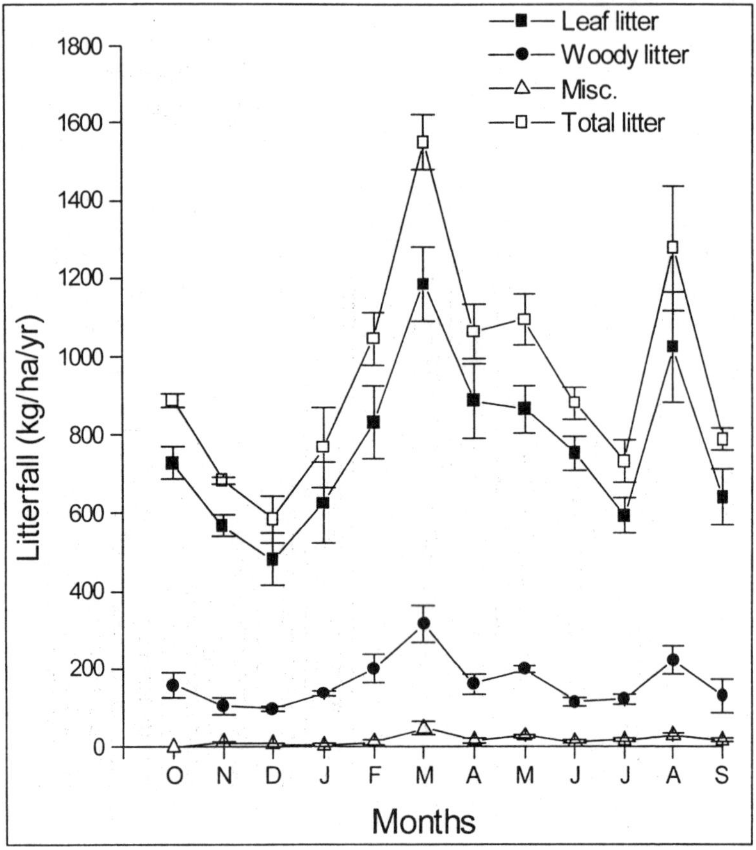

Figure 8.2: Litter Fall Pattern in the Swer Sacred Grove

to attain peak in the rainy season. Net primary productivity of fine roots in the Cherrapunji sacred grove was much lower than those found in the sacred groves of Jaintia hills (Khiewtam and Ramakrishnan 1993, Upadhaya *et al.*, 2005) most probably due to highly leached nutrient poor soil substrata, as also reported by Persson (1983). However, in both the cases the surface layer (0-10cm depth) was the zone of maximum production (Table 8.2, 8.3). The contribution of fine roots was about 50 per cent to the total (fine and coarse roots) annual production.

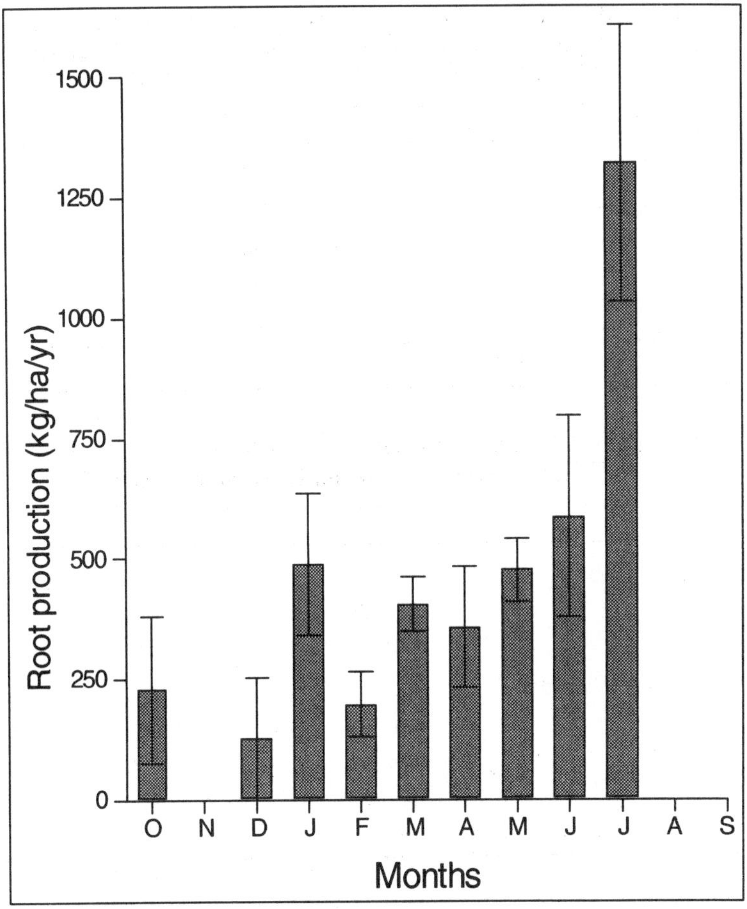

Figure 8.3: Root Production in the Swer Sacred Grove

The annual fine root production obtained in the sacred groves at Ialong and Raliang is within the range reported from dry tropical forest (Roy and Singh 1994), evergreen and deciduous tropical forests (Sundrapandian *et al.,* 1999) found in India. The observed values are however lower than re-growing pine and Oak-pine forests found in Meghalaya (John *et al.,* 2001).

Table 8.2: Productivity (kg ha^{-1} yr^{-1}) of the Fine Roots at Different Soil Depths in the Sacred Grove at Cherrapunji in Meghalaya (After Khiewtham and Ramakrishnan, 1993)

Soil Depth (cm)	*Productivity*
0	365
0-5	1093
5-10	738
10-20	325
20-30	679
Total	3200

Table 8.3: Root (Fine and coarse) Production (kg ha^{-1} yr^{-1}) in Ialong and Raliang Sacred Groves (After Upadhaya *et al.*, 2005)

	Ialong	*Raliang*
Soil depth (cm)		
0-10	6356 (58)	7073 (57)
10-20	4666 (42)	5343 (43)
Diameter class (mm)		
<2	5348 (49)	5686 (46)
2-5	2548 (23)	2348 (19)
5-10	3027 (27)	2689 (22)
10-15	99 (1)	1693 (14)
Total	11022	12416

The values in parentheses are per cent of total.

Upadhaya *et al.* (2005) reported that 67 per cent -71 per cent of the total fine root (<2 mm diameter) mass was present in the upper soil layer (0-10 cm), while the lower soil layer (10-20 cm) had only 29 per cent–33 per cent of the total fine root mass (Table 8.3). Their contribution to the total (fine and coarse) root mass in both the stands varied between 60 per cent and 66 per cent. In the sacred grove at Cherrapunji, however, only about 40 per cent of the fine root mass was present in the upper 10 cm soil layer (Khiewtam and Ramakrishnan 1993) (Table 8.4).

Table 8.4: Seasonal Distribution of Fine Root Biomass (kg ha^{-1}±SE) at Different Soil Depths in the Sacred Grove at Cherrapunji (After Khiewtham and Ramakrishnan 1993)

	Biomass (kg ha^{-1})		
	Summer	*Rainy*	*Winter*
Soil depth (cm)			
0	745±62	882±58	517±32
0-5	1376±89	2276±102	1184±65
5-10	2275±162	2912±102	2174±108
10-20	3774±190	3645±232	3319±212
20-30	3627±165	4237±222	3558±256
Root size class(mm)			
<1	2490±127	3166±188	2058±112
1-3	1941±90	2431±168	1725±90
3-6	2215±112	2419±105	1977±88
6-12	2490±108	2859±222	2406±141
>12	2661±118	3077±242	2586±155

The fine root biomass showed a marked seasonal pattern with higher values during the rainy season and lower values during spring. Low fine roots mass during spring, which marks the beginning of new shoot growth after dormant phase, could be due to their slow growth in preceding winter season and upward translocation of the reserve food. Similar seasonal trend in fine root biomass has also been reported from dry tropical and humid subtropical forest ecosystems (Srivastava *et al.*, 1986, Sundrapandian and Swamy 1996, John *et al.*, 2001).

The seasonal fluctuation was more pronounced in the upper soil layer (0-10 cm) than in the lowers layers. Rainy season being most favourable for growth after dry and cold winter, the live fraction attained peak during this season while the dead mass was more during the dry months of the year. The live biomass of both fine and coarse roots was significantly higher than the dead roots in two groves studied by Upadhaya *et al.* (2005) (Table 8.5).

Table 8.5: Seasonal Variation in Live and Dead Fractions of Fine (<2mm) and Coarse (>2mm) Root Mass (gm^{-2}) in 0-20 cm Soil Layer in Ialong and Raliang Sacred Groves (After Upadhaya *et al.*, 2005)

Sacred Grove	*Seasons*	*Fine Roots*		*Coarse Roots*	
		Live	*Dead*	*Live*	*Dead*
Ialong	Rainy	581.8±32.3	76.59±6.2	482.8±47.7	22.5±5.2
	Autumn	465.2±26.6	80.24±7.9	213.2±27.4	49.0±8.1
	Winter	435.2±26.9	87.62±6.8	242.6±48.9	23.0±6.8
	Spring	368.2±17.8	63.12±7.6	166.7±14.4	28.3±6.3
Raliang	Rainy	733.4±31.1	127.2±14.6	408.7±75.9	20.4±5.8
	Autumn	701.9±20.9	108.35±12.4	423.7±59.7	26.8±6.2
	Winter	559.8±22.5	158.55±13.3	339.0±57.6	50.6±12.2
	Spring	499.4±21.2	128.95±14.1	285.0±27.6	37.3±10.8

The seasonal variation in fine root mass reported from different forest types of India by a number of workers (Parthasarathy 1987, Srivastava *et al.*, 1986 and Sundrapandian *et al.*, 1999) has been related to soil moisture, soil temperature and root turnover. In sacred groves higher root biomass was during rainy season when climatic conditions were most favourable period for vegetative growth and availability of nutrients and water in soil was high. While during winter season when majority of plants are in dormant phase of growth, soil is relatively dry, soil temperature is low and N-mineralization rate is lower than in the rainy season, fine root mass was also low indicating the importance of these factors in fine root growth (Upadhaya *et al.*, 2005). Since the sacred grove in Meghalaya are developed on a highly leached nutrient deficient soil, high concentration of fine roots in the surface layer of the soil facilitated rapid uptake of nutrients released from the decomposing litter as well as those entering in to the soil through other sources such as throughfall and stem flow and check nutrient loss from the soil system.

8.3 Nutrient Accumulation in Litter, Fine Roots and Microbial Biomass

On the forest floor the aboveground litter was the main store house of the N followed by fine roots and microbial biomass. In case of P, litter and fine roots stored much less amount than the soil microbes which appeared to be more important pool for P than the other two biological components (Table 8.6).

Table 8.6: Mean N and P (kg ha^{-1}) Accumulation in Litter, Fine Roots and Microbial Biomass in 0-10 cm Soil Layer in the Swer Sacred Grove

Compartments	*N (kg ha^{-1})*	*P (kg ha^{-1})*
Litter	52.54±4.85	2.13±0.17
Fine roots	35.51±7.38	1.98±0.45
Microbial biomass	29.56±3.18	11.2±2.42

8.4 Nutrient Input through Litter and Fine Roots

The return of nutrients to the soil system through litter was distinctly seasonal. Its input steadily increased from the winter to spring season when forest floor is covered with a thick layer of undecomposed litter before the onset of rainy season. A smaller peak observed in Figure 8.4 is an unusual feature attributed to very heavy rain that lashed the area in August before sampling. The litter decay starts which the first few showers of monsoon rains and reaches its peak during rainy season releasing nutrients at a faster rate through mineralization process. Litter returned greater amount of N to the soil than P. The return of nutrients with fine roots, however, showed a different seasonal trend than the litter by exhibiting a peak input during rainy season and low input during winter (Figure 8.5).

Table 8.7: Annual Input (kg ha^{-1}yr^{-1}), Turnover Rate (k, yr^{-1}) and Residence Time (t, yr) of N and P in Litter and Fine Roots

Compartments	*Input (kg ha^{-1} yr^{-1})*	*Turnover Rates (k,yr^{-1})*	*Residence Time (t,yr)*
Litter			
N	113.76±6.6	2.16	0.46
P	4.67±0.33	2.2	0.45
Fine roots			
N	38.45±9.01	1.08	0.93
P	2.07±0.48	1.05	0.95

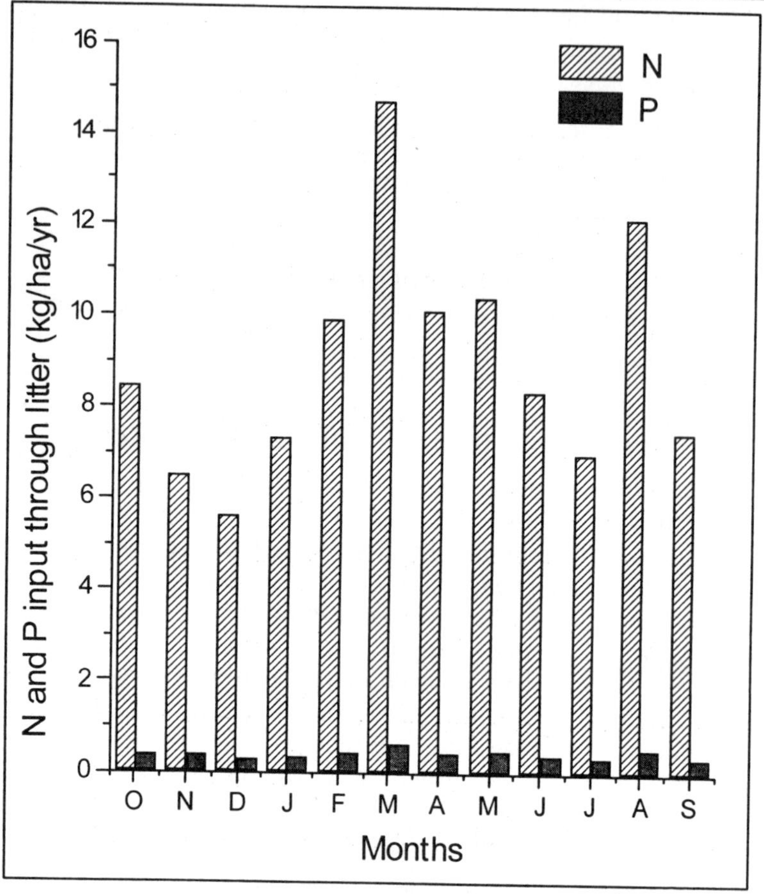

Figure 8.4: Monthly Input of N and P (kg $ha^{-1}yr^{-1}$) through Litter in Swer Sacred Grove

Annual N input through litter was three times higher than the fine roots. Litter and fine roots added relatively less amount of P into the soil (Table 8.7).

The microbes present in soil serve both as sink and source of plant available nutrients and influence their availability through the processes of mineralization and immobilization. The proportion of microbial biomass carbon (MBC) in the soil organic carbon ranged between 1.37 per cent and 2.26 per cent. The contribution of microbes

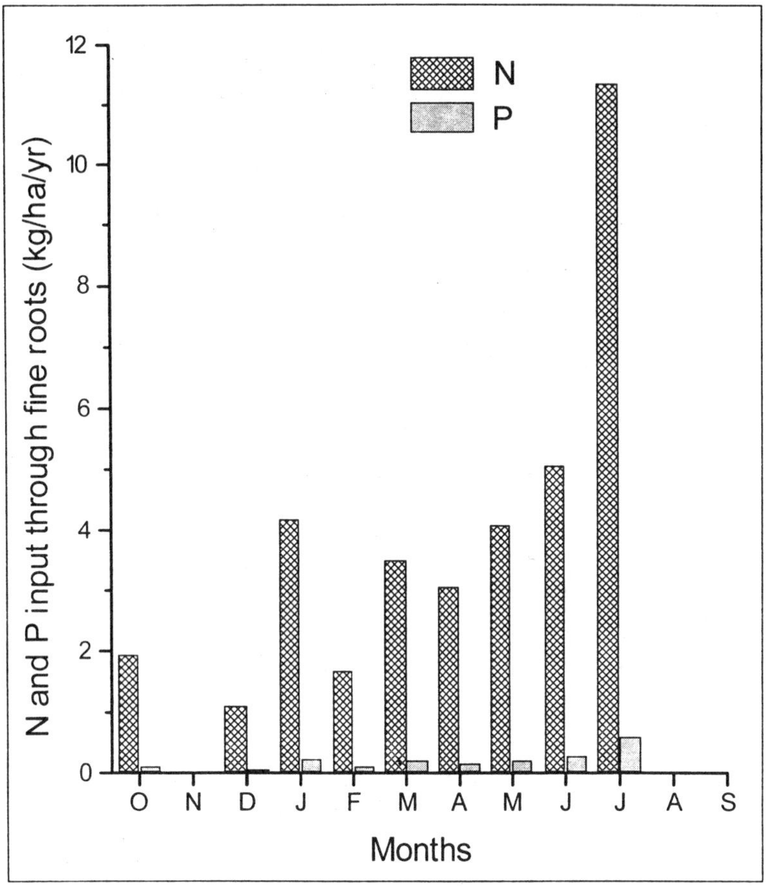

Figure 8.5: Monthly Input of N and P (kg $ha^{-1}yr^{-1}$) through Fine Roots in Swer Sacred Grove

to total soil kjeldhal nitrogen was 0.44 per cent to 0.47 per cent. Their contribution to the soil P pool was still less. Thus the microbes accumulated a small fraction of total nutrients present in the soil. They accumulated nutrients during dry winter and spring seasons, a process commonly called as immobilization, and released them during rainy season when nutrients were in high demand by the vegetation. This results into a distinct seasonal variation in the concentrations of microbial biomass carbon, nitrogen and phosphorus in soil (Table 8.8).

Table 8.8: Seasonal Variation in Microbial Biomass Carbon (MBC), Microbial Biomass Nitrogen (MBN) and Microbial Biomass Phosphorus (MBP) in the Soil of Swer Sacred Grove

Seasons	*MBC (μg g^{-1})*	*MBN (μg g^{-1})*	*MBP (μg g^{-1})*
Autumn	424.98±34.36	24.03±1.19	12.98±2.72
Winter	610.97±113.20	59.21±12.41	17.81±3.05
Spring	620.52±104.19	42.18±2.17	14.82±2.46
Rainy	341.57±39.52	23.63±1.5	7.85±1.64

8.5 Decomposition of Leaf Litter and Fine Roots

Turnover rates of N and P through litter was much faster than fine roots which decomposed at a slower rate than litter (Table 8.7). A study of decomposition of leaf litter of three dominant tree species and fine roots in Swer sacred grove revealed that the former decomposed at a faster rate than the latter (Figure 8.6). About 44 per cent–62 per cent of the initial amount leaf litter got decayed during one year period whereas only 39 per cent of fine roots were decomposed during the same period. N release from decomposing leaf litter was almost double than those from decaying fine roots. P release from decomposing leaf and root litter was slower than N release.

Table 8.9: Decay Constant (k) and N and P Mineralization Constants (k_N, k_P) of Tree Leaf Litter and Root Litter

Litter Types	*Dry Matter (k, $year^{-1}$)*	*N Mineralization (k_N, $year^{-1}$)*	*P Mineralization (k_P, $year^{-1}$)*
Myrica leaf litter	0.53	0. 68	0. 62
Rhododendron leaf litter	0.82	1. 12	0. 84
Neolitsea leaf litter	0.89	1. 39	0. 88
Root	0.39	0. 73	0. 63

8.6 N and P Mineralization

The litter of above and below ground origin is decomposed by the activities of large variety of soil microbes present in the soil. The whole process of decomposition is completed in two major steps namely, humification-transformation of freshly fallen litter into

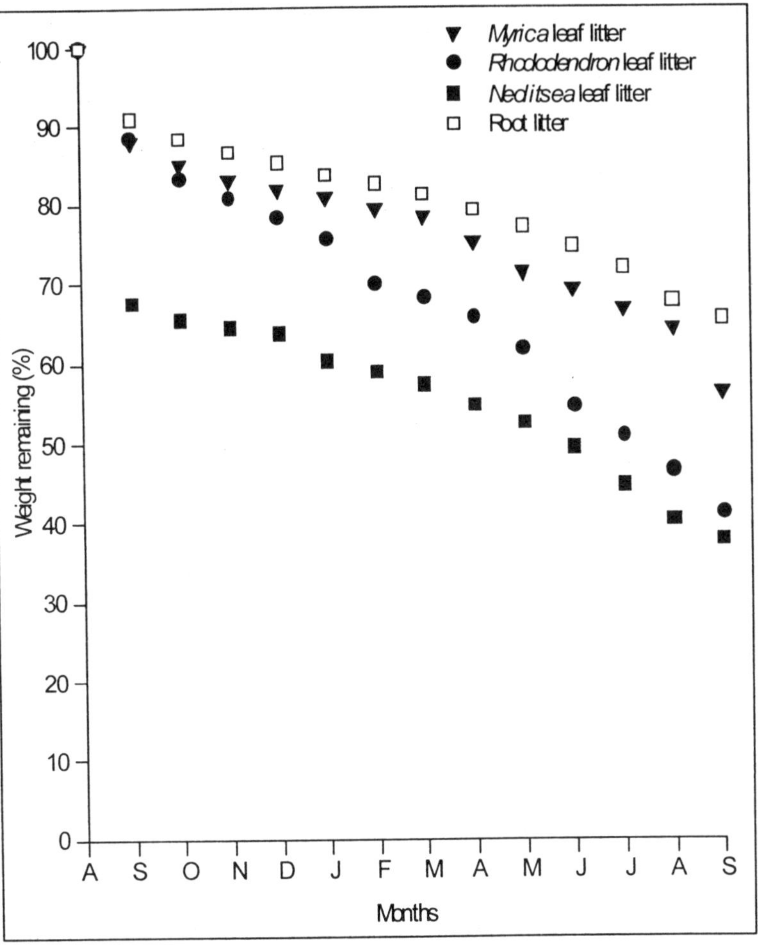

Figure 8.6: Decay Pattern of Tree Leaf Litter and Fine Roots Litter in Swer Sacred Grove

humus and mineralization-conversion of humus into soluble inorganic nutrients. The rate of decomposition depends on several factors; important among them are climate and soil conditions and quality and quantity of organic resource present on the forest floor. In the tropics where climate is distinctly seasonal, decomposition process follows a seasonal pattern with fast rate during wet-warm rainy season and slow rate during dry-cold winter season. Since all studied sacred groves are located in humid subtropical climate zone, both net N and P mineralization rates showed a sharp seasonal variation (Figure 8.7). The mineralization was at its peak during rainy season thereafter it steadily decreased attaining lowest value during winter season. Thus periods of active plant growth and nutrient release in the soil were closely synchronized within the

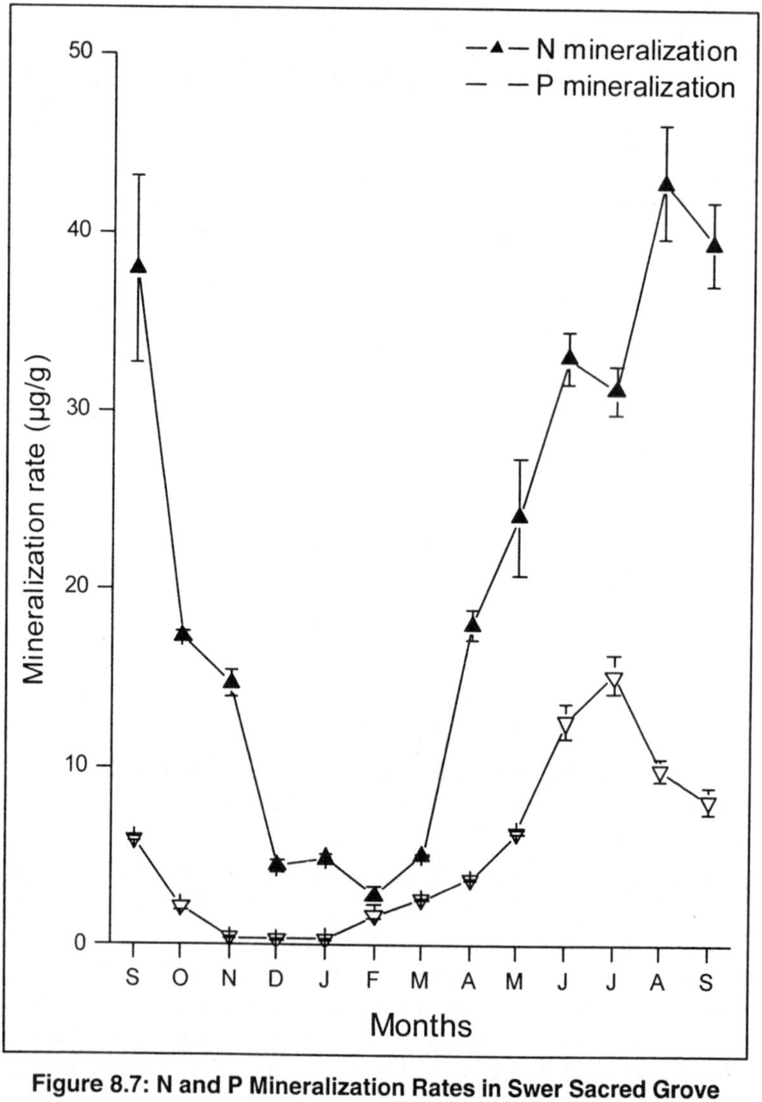

Figure 8.7: N and P Mineralization Rates in Swer Sacred Grove

community. Similarly when the nutrient requirement of majority of species in the community is at minimum level during winter season nutrient supply through mineralization was also at the lowest level (Figure 8.7).

An over view of nutrient (N and P) dynamics through litter, fine roots and microbial biomass in the Swer sacred grove is presented in Figure 8.8. Annual input to soil through litter was 11371 kg ha^{-1} dry matter, 114 kg ha^{-1} N and 5 P kg ha^{-1} while the corresponding input from fine roots was 4367 kg ha^{-1} dry matter, 38.4 kg ha^{-1} N and

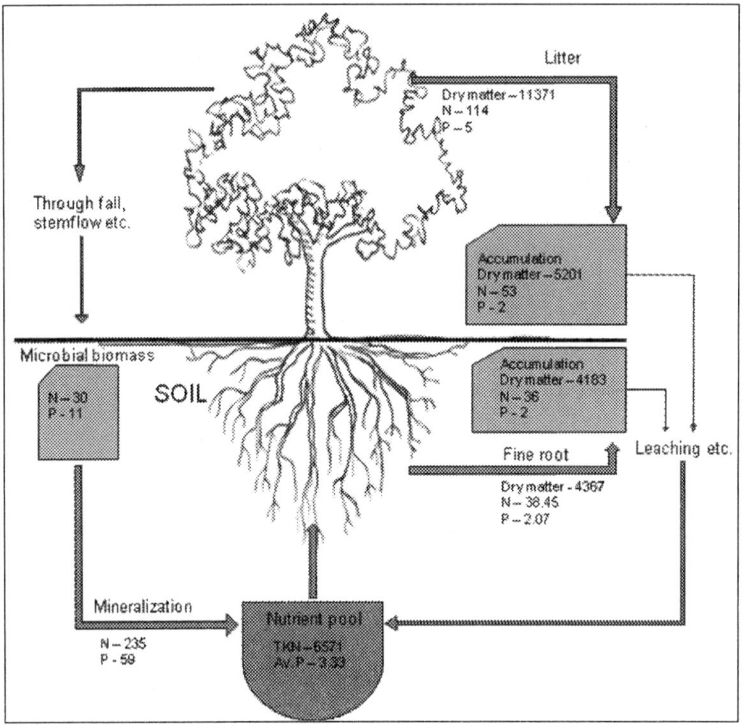

Figure 8.8: Nutrient Dynamics through Litter, Fine Roots and Microbial Biomass in Swer Sacred Grove

The values in the boxes are accumulation (kg ha^{-1}) while those on arrows are rate (kg ha^{-1} yr^{-1}). The values for fine roots (≤2 mm diameter), microbial biomass and soil nutrient pool are for upper (0-10 cm) layer of soil profile. TKN- Total kjeldhal nitrogen, Av.P- available phosphorus.

2.1 kg ha^{-1} P kg ha^{-1}. Litter was the major store house of organic matter and N on the forest floor followed by fine roots. However, relatively greater amount of P was present in the microbial biomass, which was released at much slower rate than N during mineralization process; the mean net N and P mineralization rate being 234.53 kg ha^{-1} yr^{-1} and 59.11 kg ha^{-1} yr^{-1} respectively.

Thus limited amount of data presented in the foregoing pages reveals that litter, fine roots and microbes present on the forest floor and in the upper layer of mineral soil in sacred groves play important role in storage and circulation of large amount of essential nutrients like nitrogen and phosphorus. Therefore, any anthropogenic disturbance which destroyes the vegetation is likely to disrupt the nutrient circulation and adversely affect other proceses in the wihin the ecosystem. Some of the effects of anthropogenic disturbance on the ecological processes in the sacred grove are discussed in the next chapter.

Chapter 9

Impact of Anthropogenic Disturbance and Conservation of Sacred Groves

9.1 Impact of Anthropogenic Disturbance

Most of the world's primary forests are subjected anthropogenic disturbances to meet various human needs. These disturbances have been reported to alter species composition, structural organization and dynamics of community, which in turn influences almost all ecosystem functions. Disturbance that involve single factor such as selective logging limited to small area brings relatively minor changes in structural attributes of the forest and often enhance species diversity. But frequent low intensity disturbance such as grazing, browsing and fuel wood extraction and surface fire may cause greater damage to forest structure and functions. There are two views regarding the effect of disturbance on species diversity. One hypothesis is that ecosystems with greater species diversity are more resilient to environmental disturbance and some species would compensate for those members of the community that are reduced or eliminated by the disturbance (Pimm 1984, Ehrlich and Wilson 1991, Lawton and Brown 1994). Contrary to this view Connell (1978), based on his studies on rain forest, proposed that the tree species diversity would be greatest where disturbances are moderate in intensity and frequency. Similar view was expressed by Collins *et al.* (1995) who argued that species richness should be highest at intermediate frequencies of disturbance when condition favours competitive species and those that tolerate disturbance to co-exist.

In a closed canopy forest like sacred groves damage caused to the trees either by natural agencies such as wind and hail storms, land slips etc. or by human activities such as selective tree cutting may evidently lead to the formation of gaps of different sizes and shapes in the forest canopy. Such openings significantly alter the microclimate and microsite heterogeneity on the forest floor to the extent that it may affect community dynamics and play important role in maintaining species diversity in the forest. In course of time, the gaps are colonized by a variety of plant species making the forest a mosaic of plant communities representing different developmental stages of the forest (Whitmore 1975).

The erosion of religious beliefs and taboos associated with the sacred groves all over the country has exposed them to varied types of anthropogenic disturbances. As a result illicit cutting of trees and grazing, conversion of sacred groves into plantations and encroachment for developmental activities and urbanization and many more human activities has become common in the sacred groves found in different parts of the country, and very few of them represent the virgin forest of the area. The damage caused to the sacred grove vegetation and its impact on the entire ecosystem depends on the type, intensity and frequency of disturbance that vary from place to place.

In Meghalaya also sacred groves are not free from the human disturbances, rather most of them are affected by human activities such as shifting cultivation, timber and firewood extraction and developmental and urbanization activities. They are being encroached and converted into village reserve forests to meet timber and fuel wood needs of the villagers residing close to the grove. At some places the vegetation in the periphery of sacred forest is burnt annually in dry winter to promote herbage growth for cattle grazing during ensuing spring and rainy seasons.

Several comparative studies have been conducted in the protected and adjoining disturbed stands of different sacred groves of Meghalaya to evaluate the effect of prevailing anthropogenic disturbances on vegetation and soil characteristics. Findings of these studies suggest that disturbance results in to (*i*) a species-poor community, (*ii*) alter species composition with an increase in the species typical of open habitats, (*iii*) disappearance of many mature phase forest species and colonization of forest floor by new species, (*iv*) impairment of tree regeneration process and development of unstable population structure and (*v*) change in soil properties and fine root dynamics and nutrient mineralization rate. Major findings on the above aspects are summarized in the following pages.

(*i*) Soil Properties

A study carried out by Upadhaya *et al.* (2005) on the soil properties and fine root dynamics and nutrient mineralization rate in protected (undisturbed) and unprotected (disturbed) stands of two sacred groves revealed that the soil in the protected stand was more acidic than the unprotected stands and acidity increased with the increase in soil depth (Table 9.1). The Soil Organic Carbon (SOC), Total Kjeldhal Nitrogen (TKN) and available-P concentration was significantly higher in the protected stands than the unprotected stands. The surface layer (0-10cm) had significantly higher concentration of SOC, TKN and available-P than the subsurface (10-20cm) soil layer.

(*ii*) Species Richness and Diversity

In another study Upadhaya *et al.* (2008) reported that as a response to disturbance, the woody species (≥ 5cm dbh) richness got significantly reduced in the three groves (Table 9.2). The lianas which were few in numbers in all studied forests were completely eliminated from the disturbed stands. Several endemic species of northeast India such as *Ardisia griffithii, Camellia cauduca, Carpinus vareca, Erythroxylon kunthianum* etc. and threatened category of species such as *Fraxinus floribunda, Hedera helix, Lindernia latifolia, Mangletia insignis* etc., which were present in the undisturbed stands were either poorly represented or completely absent in the disturbed stands indicating substantial loss of taxonomically and ecologically important species from the forest. The movement of humans and grazing animals and fire incidence during dry winter season was the main cause of development of species-poor community in the disturbed stands. The Shannon's diversity index computed using density of woody species varied from 2.02 in Nongkrem disturbed stand to 3.26 in Nonglang undisturbed stand. The diversity index was invariably low and Simpson's dominance index was high in the disturbed stands (Table 9.2).The density of woody species also showed a marked reduction from the undisturbed to the disturbed stands in two groves, however, in Nongkrem sacred grove there was a marginal increase due to abundant growth of shade-intolerant pioneer tree species *Pinus kesiya* in forest openings created by disturbance.

Table 9.1: Physico-chemical Properties of Soils of Protected and Unprotected Stands of the Sacred Groves at Ialong and Raliang, Jaintia Hills, Meghalaya Values of SOC, TKN and available-P are the means of four seasons across the year (After Upadhaya *et al.*, 2005)

Soil Properties	*Soil Depth (cm)*	*Ialong*		*Raliang*	
		Protected	*Unprotected*	*Protected*	*Unprotected*
Sand (Per cent)	0-10	84.6	84.4	58.8	81.2
	10-20	81.4	77.6	37.5	73.2
Clay (Per cent)	0-10	4.5	4.5	16.4	6.0
	10-20	7.2	8.0	26.7	13.2
pH	0-10	4.8	5.1	4.6	4.9
	10-20	4.5	4.8	4.4	4.6
SOC (Per cent)	0-10	4.9	4.1	4.9	4.4
	10-20	4.4	3.8	4.3	3.9
TKN (Per cent)	0-10	0.45	0.38	0.47	0.38
	10-20	0.36	0.32	0.42	0.31
Avail-P	0-10	15.35	14.23	17.90	13.10
	10-20	12.68	10.08	13.32	9.47

Effect of anthropogenic disturbance on the understorey plant diversity in two sacred groves of Meghalaya was studied by Upadhaya *et al.* (2006). The understorey had predominance of members of Asteraceae, Fabaceae, Commelinaceae, Zingiberaceae, Orchidaceae and Rubiaceae. In this respect the community was similar to the forest in the Western Ghats and other rain forests in the tropics. They observed an increase in species richness and diversity of both shrubs and herbs in the disturbed stand than the adjoining undisturbed stands. The density of shrubs was also high in the disturbed stand, but contrary to this herbs' density was high in the undisturbed grove (Table 9.3). The increase in species richness of the forest understorey was due to colonization by fast growing secondary species like *Artemesia nilagirica, Eupatorium adenophorum, Lantana camara, Melastoma nepalensis* and *Sida rhombifolia* among the shrubs and *Drymeria cordata, Hedyotis tenelliflora, H. unicinella, Potentilla fulgens, Plantago erosa* and *Polygonum* species among the herbs in the open spaces created by selective tree cutting.

Table 9.2: Diversity of Woody (≥ 5cm dbh) Species and Other Community Characteristics in Three Undisturbed and Disturbed Stands of Sacred Groves of Meghalaya (Upadhaya *et al.*, 2008)

Variables	*Swer*		*Nonglang*		*Nongkrem*	
	**U*	*D*	*U*	*D*	*U*	*D*
Area sampled (ha)	0.5	0.5	0.5	0.5	0.5	0.5
Species richness	60	26	56	24	32	24
No. of genera	45	22	44	19	20	28
Shannon diversity index	2.82	2.66	3.26	2.52	2.40	2.02
Simpson dominance index	0.120	0.105	0.061	0.126	0.147	0.190

*U: Undistributed; D: Distributed.

(*iii*) Tree Population Structure

Upadhaya *et al.* (2004) studied tree population structure by analyzing density-diameter distribution pattern in the undisturbed and disturbed stands of two sacred groves in Meghalaya. In all the four stands density-diameter yielded a reverse J-shaped curve indicating preponderance of young individuals in the population. Similar age structure of tree population was observed when densities of seedling, sapling and adult trees were taken in to consideration (Table 9.4). The population density declined during successive stages of tree growth from seedling to adult stage, but the decline was much sharp from seedling to sapling stage. This indicated that though there was large recruitment of seedlings in the forest, their survival percentage was low due to heavy mortality caused both by unfavourable physical and biological environmental factors. Mild anthropogenic disturbance has little or no effect on the tree population structure. Similar tree population structure has been reported from the forests at Costa Rica (Nadkarni *et al.*, 1995), Brazalian Amzon (Campbell *et al.*, 1992) and Eastern Ghats (Kadavul and Parthasarathy 1999).

Table 9.3: Species Diversity and Other Characteristics of Understorey Plant Community in the Undisturbed and Disturbed Sacred Grove Stands at Ialong and Raliang in Jaintia Hills, Meghalaya (after Upadhaya *et al.*, 2006)

Parameters	*Ialong*		*Raliang*	
	Undisturbed	*Disturbed*	*Undisturbed*	*Disturbed*
Shrub				
Density ha^{-1}	6428	10680	8064	11038
Species richness	49	61	46	59
Number of families	30	32	28	31
Number of genera	39	44	37	43
Shannon diversity index	3.40	3.76	3.04	3.72
Pielou evenness index	0.61	0.63	0.55	0.63
Simpson dominance index	0.048	0.029	0.078	0.031
Herb				
Density ($100m^2$)	2474	2078	1997	1887
Species richness	52	56	45	51
Number of families	23	30	20	28
Number of genera	43	50	39	46
Shannon diversity index	3.26	3.59	3.30	3.45
Pielou evenness index	0.57	0.62	0.60	0.61
Simpson dominance index	0.067	0.036	0.049	0.043

(*iv*) Tree Regeneration

Upadhaya *et al.* (2008) assessed the influence of disturbance on regeneration of woody species based on seedling and sapling population density in three undisturbed and disturbed sacred groves in Meghalaya. The number of species showing good regeneration was higher in the undisturbed than the disturbed stands of two groves. Similarly the number of species showing no regeneration was also high in the undisturbed stands. Fairly large number of species showed good regeneration in the disturbed stands (Table 9.5) suggesting that altered forest microenvironment favoured their regeneration. Dominant species exhibited good to fair regeneration while less abundant species were either poorly regenerating or not regenerating.

Table 9.4: Age Structure of Tree Population (Density 100m²) in Disturbed and Undisturbed Stands of Three Sacred Groves in Meghalaya (After Upadhaya *et al.*, 2004)

Density	*Swer*		*Nonglang*		*Nongkrem*	
	**U*	*D*	*U*	*D*	*U*	*D*
Seedlings	21150	10750	9475	3975	5375	5075
Saplings	1880	1692	1508	748	1148	1032
Adult trees	13	3	10	3	9	10

*U: Undistributed; D: Distributed.

At community level regeneration appeared satisfactory, but some species especially those of climbers/lianas showed no sign of regeneration. There was abundance of individuals in smaller diameter classes, while trees in the large diameter classes were usually absent in the disturbed stands. The density of seedlings, saplings and adult trees indicated much greater loss of individuals during conversion of seedlings to saplings which happens to be a vulnerable stage in the life cycle of the species than the period during transformation of saplings to adult trees. High mortality and low growth rate are typical of juvenile rainforest trees due to low light level in understorey, exposure to physical and biotic disturbances and short-term water deficit. Abundant seedling recruitment in the undisturbed groves was attributed to presence of seed bearing parent trees and low level of disturbance. On the other hand, in the disturbed stand less number of seed bearing trees and concurrent disturbance were responsible for low seedling density. The density was fairly uniformly distributed among different species in the undisturbed forest, it was lop sided in favour of *Rhododendron arboreum*, and *Myrica esculenta* at Swer, *Pinus kesiya* and *Eurya acuminata* at Nonglang and *Myrica esculenta* and *Pinus kesiya* at Nongkrem disturbed stands. They performed better in the disturbed stands.

Table 9.5: Number of Species Regenerating in Undisturbed and Disturbed Stands of Three Sacred Groves of Meghalaya (After Upadhaya *et al.*, 2008)

Sacred Groves	*Stands*	*Regeneration Categories*				
		Good	*Fair*	*Poor*	*None*	*New*
Swer	Undisturbed	18	1	6	35	1
	Disturbed	15	0	3	8	3
Nonglang	Undisturbed	23	5	4	24	1
	Disturbed	15	0	4	5	6
Nongkrem	Undisturbed	12	5	3	12	5
	Disturbed	14	5	1	4	1

Barik *et al.* (1996a) investigated seed production, dispersal and germination of three broad-leaved tree species *viz. Lithocarpus dealbatus, Quercus griffithii and Schima khasiana* in the undisturbed, mildly disturbed and highly disturbed sacred groves at upper elevation of Meghalaya and found that seed production in *L. dealbatus* and *S. khasiana* increased with increasing disturbance. The dispersal distance for *Lithocarpus* and *Quercus* seeds from the trees was greater in the disturbed groves than in the undisturbed grove. Seed predation decreased and germination increased with distance from the parent tree in all three stands. Germination of *Lithocarpus* and *Quercus* was better in the undisturbed stand than in the disturbed stands, while that of *S. khasiana* was better in the disturbed stands.

Barik *et al.* (1996b) further studied recruitment and survival of *L. dealbatus, Q. griffithii, Rhododendron arboretum* and *S. khasiana* in the three stands. Maximum recruitment of S. *khasiana seedlings* was in highly disturbed stand, while *L. dealbatus* seedlings were abundant in the undisturbed and mildly disturbed stands. Seedlings of *L. dealbatus, Q. griffithii, R. arboretum* were equally successful in all the stands, *S. khasiana* showed better survival in highly disturbed stand than in the mildly disturbed and undisturbed stands. In *L. dealbatus, Q. griffithii* and *S. khasiana* age-specific mortality rate was significantly affected by photosynthetically active radiation, relative humidity, air temperature, soil temperature, soil moisture and thickness of litter on the forest floor, while in *R. arboretum* mortality was not related to any of the above factors. On the basis of seedling recruitment and survival pattern, they concluded that *L. dealbatus, Q. griffithii, R. arboretum* were successful in both disturbed and undisturbed stands, while *Schima khasiana* seedlings showed better survival and establishment in the disturbed stand.

(*v*) Fine and Coarse Root Dynamics and N-mineralization

The accumulation of fine (<2mm diameter) and coarse root (2-15mm diameter) mass in the protected stand was significantly ($P<0.01$) higher than their corresponding unprotected stands of two groves in Jaintia hills. The protected stand had greater live as well as dead fine root mass than the unprotected stand. In coarse roots also live fraction was greater in the protected stand but in case of dead fraction the trend was not consistent (Table 9.6). Most of the fine roots (67-71 per cent) were concentrated in the upper soil layer in all the four stands. The high concentration of fine roots in the surface layer was attributed to better aeration, and high organic matter, nitrogen and phosphorous contents in this layer (John *et al.*, 2001). This is an important feature of humid tropical ecosystem where fine roots play crucial role in storage and rapid recycling of nutrients. Thus any damage caused to these surface feeding root systems is likely to increase nutrient losses from the system and adversely affect water and nutrient absorption by the vegetation.

Root Production

The total annual root production was higher in the protected stands (Raliang 783 gm^{-2} and Ialong 642 $g\ m^{-2}$) than their corresponding unprotected stands (Raliang 624 $g\ m^{-2}$ and Ialong 520 $g\ m^{-2}$) (Table 9.7). The surface layer accounted for 62 per cent–68 per cent of the total root production. A significant reduction in the fine root mass and annual production in the disturbed stands was attributed to the change in species composition (Upadhaya *et al.*, 2005) and alterations in the soil physical properties due to cattle trampling and low N and P and organic matter contents in the soil.

N-mineralization

The results showed that N mineralization rate was significantly higher in the undisturbed stands than the disturbed stands; the rate being significantly higher in the upper soil layer rate than the lower layer (Table 9.8). Disturbance in the forests brought about reduction in both mineral-N pool as well as N-mineralization rate. This was mediated through change in species composition and decline in species dominance (tree basal cover) in the community. Thus the adverse effect of aboveground changes in the community composition due to anthropogenic disturbances of mild intensity was clearly evident on the soil physico-chemical properties and fine root dynamics in the sacred groves.

Table 9.6: Mean Live (L) and Dead (D) Dry Mass of Fine (<2mm) and Coarse (2-15mm) Roots (gm^{-2}) in Protected and Unprotected Sacred Groves at Ialong and Raliang, Jaintia Hills, Meghalaya (After Upadhaya *et al.*, 2005)

Stands	*Depth (cm)*	*Fine Root*			*Coarse Root*		
		L	*D*	*Total*	*L*	*D*	*Total*
Ialong							
Protected	0-10	319.99	43.38	363.37	135.02	19.17	154.19
	10-20	143.43	33.50	176.93	141.23	11.49	152.72
Unprotected	0-10	233.94	62.75	296.68	113.20	17.03	130.23
	10-20	90.53	34.41	124.95	104.8	12.30	117.10
Raliang							
Protected	0-10	444.19	74.35	518.54	178.67	17.54	196.21
	10-20	179.43	56.37	235.81	184.60	16.25	200.85
Unprotected	0-10	258.10	71.43	329.53	152.17	34.74	186.91
	10-20	92.05	44.04	136.09	92.48	25.93	118.41

Table 9.7: Root Production ($g\ m^{-2}\ yr^{-1}$) in the Protected and Unprotected Sacred Groves at Ialong and Raliang, Jaintia Hills, Meghalaya (The values in parentheses are per cent of total) (After Upadhaya *et al.*, 2005)

Categories	*Ialong*		*Raliang*	
	Protected	*Unprotected*	*Protected*	*Unprotected*
Diameter class (mm)				
<2	425.33 (66)	338.51 (65)	426.10 (57)	356.60 (69)
>2-15	216.72 (34)	285.96 (35)	356.62 (43)	163.37 (31)
Soil depth (cm)				
0-10	418.33 (65)	347.82 (62)	513.17 (68)	337.35 (65)
10-20	223.72 (35)	276.65 (38)	269.55 (32)	182.62 (35)

Table 9.8: Average N-mineralization Rate (mg $g^{-1}month^{-1}$) in the Protected and Unprotected Sacred Groves at Ialong and Raliang in Jaintia Hills, Meghalaya (After Upadhaya *et al.*, 2005)

Sacred Grove	*Soil Depth (cm)*	*Ammonification*	*Nitrification*	*Mineralization*
Ialong				
Protected	0-10	8.31±1.70	5.82±1.17	14.13
	10-20	6.31±1.51	4.75±1.17	11.06
Unprotected	0-10	5.43±0.96	6.01±1.16	11.44
	10-20	4.59±0.77	4.61±1.20	9.20
Raliang				
Protected	0-10	7.92±1.57	5.84±1.14	13.76
	10-20	6.68±1.58	4.47±0.97	11.15
Unprotected	0-10	6.09±1.03	5.77±1.17	11.86
	10-20	4.93±0.96	4.62±1.05	9.55

±SEM (n=12).

9.2 Conservation of Sacred Grove

In the past protection and conservation of the sacred grove were closely linked with the religious beliefs and traditional culture of the indigenous tribal communities all over the world (Malhotra *et al.*, 1997). This cultural heritage survived through generations, but unfortunately the religious beliefs and taboos that were in the centre of sacred grove preservation are now fast eroding due to change in social set up, increased consumerism for better living and change in the belief system of the people attributable to modern system of education (Rao, 1996). Therefore it is extremely difficult, if not impossible, to protect sacred groves indefinitely only on the basis of religious beliefs. There is a need to revitalize the age-old ethos of conservation culture of traditional societies by supplementing with scientific knowledge about the crucial role these forest patches play in conserving biodiversity and providing large number of valuable goods and vital ecological services to the people living around the sacred groves. Therefore urgent interventions are required for the conservation of these forest patches.

The steps taken by government agency(s) for the conservation of this vital natural resource often do not yield desired result on account of non-involvement of local people in decision making process and implementation of the programmes. Therefore, to make any conservation programme successful participation of the local people especially the youth is essential. This is possible by educating them about the importance of conservation of these forest patches by highlighting their direct and indirect and short and long-term benefits to both local community and to the area at large. At the same time there is need to train local people about the sustainable use of plant resources such as medicinal, food, fuel wood, timber and other economically important plants needed in day-to-day requirement to check their over exploitation.

Jayarajan (2004) in his study on sacred groves of north Malabar suggested several measures to check degradation and conservation of sacred groves, many of them are relevant for the conservation of sacred groves found elsewhere in the country.

1. The faith in sacredness of groves should be respected and supplemented with ecological knowledge.
2. The ownership of sacred groves should be transferred to the communities from the individuals.
3. The state government should declare sacred groves as heritage sites and provide financial support for their up keep including fencing.
4. Resurvey of sacred groves in different states to assess their present condition and recover the encroached areas if possible.
5. Conservation of intact groves and restoration of degraded ones.
6. Awareness campaign should be organized for the local people focusing the need of protection of the groves.
7. A buffer zone between sacred groves and the surrounding man-impacted landscape elements may be created to check the anthropogenic disturbances and provide conditions for regrowth of the forest.

The study highlighted the need for building local ecological knowledge and incorporating it in strengthening the management of sacred grove institution.

In Meghalaya Regional Centre, National Afforestation and Ecodevelopment Board, Ministry of Environment and Forests, Shillong has taken various initiatives for the conservation of sacred groves. In one of the workshops several suggestions were made for sacred grove protection by grass-root level functionaries associated with its management in the state. The important points that emerged in the workshop are as under:

- ✰ State forest department may formulate management plans for conservation of medicinal, rare and endangered species which are mostly found in the sacred groves. The lower functionaries of the department should be involved in identifying traditional medicinal flora and creating awareness among the people about the commercial importance of such species.
- ✰ The government may organize local communities into small cooperatives societies for scientific collection of the medicinal plants.
- ✰ The local people may serve as watchdogs for the unauthorized collection of such species by outside agencies.
- ✰ The policy/rules should be formulated and strictly implemented for the conservation of groves.

Besides these, the following steps may help in the conservation of sacred forests and their biological contents.

1. The endemic and threatened plant species confined to sacred groves should be identified and steps should be taken to conserve these species. The well-preserved sacred groves should be immediately brought under protected area network to ensure the protection of such species and their habitats.
2. A sacred grove conservation programme should be initiated keeping the traditional administrative bodies at the centre of the programme. There is a need to convert the traditional belief of the tribal people into an effective conservational strategy. The scientific basis and conservational values behind the belief need to be explained to the villagers. Programmes may be organized

to bring awareness among all sections of the community including school children.

3. Adequate funding should be given to the village *dorbars* to create tree plantations to meet the biomass needs of the villagers so that anthropogenic pressure on the sacred grove may be reduced.
4. Ecological studies of threatened species need to be undertaken to understand their environmental requirements. This will help in working out suitable strategies for their propagation and continued existence.
5. Development of nursery technique for the native species confined to the sacred groves should be undertaken.
6. The degraded sacred groves should be immediately restored or regenerated using appropriate technology. Efforts should be made to utilize the traditional knowledge in regenerating those native species for which technology may not be available.
7. The Traditional heads/villagers who engage themselves in protection and conservation of biodiversity-rich areas need to be rewarded. The most viable strategy for the protection of sacred groves seems to lie in linking their conservation with economic benefits. Rewards and incentives to the people may go a long way in achieving the goal.
8. There is need to highlight the ecological services rendered by sacred groves and people should be made to realize that the conservation of groves is crucial for their sustenance.

At national level Indira Gandhi Manav Sangrahalaya, Bhopal in collaboration with many academic institutions and NGOs has undertaken a number of activities for the protection and conservation of sacred groves (refer Malhotra *et al.*, 2007). Some of the initiatives and activities undertaken by Indira Gandhi Manav Sangrahalaya are as follows:

1. Replicas of sacred groves of Arunachal Pradesh, Chattisgarh, Kerala, Maharashtra, Manipur, Meghalaya, Rajasthan, Tamil Nadu and West Bengal were created at Bhopal to serve as living nurseries of ancestral and

community identity, and purity and longevity in community habitats.

2. Indoor exhibitions were organized using photographs, maps and charts etc. in different parts of the country.
3. Traveling exhibition was organized in different parts of the country for greater interaction with local people and organizations to learn more about sacred groves of the country.
4. Sacred grove festivals were organized involving grass-root functionaries associated with sacred groves from 15 states, foresters, scientists and media persons to discuss various aspects related to sacred groves.
5. Publication of literature on sacred groves.

A thematic group on Culture and Biodiversity was constituted under National Biodiversity Strategy and Action Plan, executed by Ministry of Environment and Forests, Government of India, to chalk out future strategy and action plan for sacred groves at the national level. The group under the coordinatorship of Prof. K.C.Malhotra made following recommendations for strengthening the sacred grove (SG) institution of the country.

1. An inventory of SGs covering different aspects such as religious, cultural, social, economic, ecological etc. should be prepared for all the states of the country.
2. Mapping of the nature and extent of vegetation cover in SGs throughout the country should be done.
3. The Ministry of Environment and Forests, Government of India should develop a policy document on SGs.
4. Campaign should be undertaken to strengthen SGs which are facing threats of degradation in different parts of the country. This should aim at people who live in vicinity of SGs. They should be made aware of the ecological functions performed by SGs and motivate them to protect it.

Appendices

Table 1: Sacred Groves of Meghalaya Studied in Detail by Ecology Research Group, Department of Botany, NEHU, Shillong

Name of the Sacred Grove	*Geographical Location*	*Workers*
Mawphlang, East Khasi hills	25°34' N latitude, 91°56'E longitude, 1825 m altitude	Khan ML, 1986 Barik SK, 1992 Rao P, 1992 Das AK, Boral L, Tripathi RS and Pandey HN, 1997
Shorarim, East Khasi hills	25°21' N latitude, 91°44' E longitude, 1665-1695m altitude	Nongbri S, 2007
Swer, East Khasi hills	25°25' N latitude, 91°47' E longitude, 1910-1975m altitude	Mishra BP, Tripathi OP, Tripathi RS and Pandey HN, 2004 Nongbri S, 2007 Kamei J, 2007
Cherrapunji, East Khasi hills	25°15' N latitude, 91°45' E longitude, 1313m altitude	Khiewtam RS, 1986
Khloo Blai, Jaintia hills	25°26' N latitude, 92°12' E longitude, 1200-1300m altitude	Jamir AS, 2000 Odyuo N, 2004
Khloo Poh Lyngdoh, Jaintia hills	25°26' N latitude, 92°12' E longitude, 1200-1300m altitude	Jamir AS, 2000 Odyuo N, 2004, Law P, 2002
Khloo Paiu Ram Pyrthai, Jaintia hills	25°26' N latitude, 92°12' E longitude, 1200-1300m altitude	Jamir AS, 2000 Odyuo N, 2004
Khloo Langdoh, Jaintia hills 1300m altitude	25°30' N latitude, 92°28' E longitude,	Upadhaya K, 2002 Law P, 2002
Urkhla, Jaintia hills	25°28' N latitude, 92°16' E longitude, 1350m altitude	Upadhaya K, 2002 Law P, 2002
Nonglang, East Khasi hills	25°21.8' N latitude, 92°27.9' E longitude, 1541m altitude	Upadhaya K, Barik SK, Pandey HN and Tripathi OP, 2008
Nongkrem, East Khasi hills	25°25.54' N latitude, 91°52.49'E longitude, 1781m altitude	Upadhaya K, Barik SK, Pandey HN and Tripathi OP, 2008

Table 2: Distribution of Sacred Groves in India (After Malhotra 1998, Malhotra *et al.*, 2001, Malhotra *et al.*, 2007, Khan *et al.*, 2008)

Location	*Number*	*Districts*	*Area*
Andhra Pradesh	800	–	–
Andhra Pradesh	750	23	–
Arunachal Pradesh	159	6	–
Assam	40	1	–
Gujarat	29	1	1 acre to 2 sq. km
Haryana	248	18	–
Himachal Pradesh	11	–	–
Karnataka (Coorg)	1214	–	2407 ha
Kerala	2000	–	500 ha
Madhya Pradesh	275	–	–
Maharashtra	1600	–	–
Maharashtra	483	10	3570 ha
Maharashtra	250	1	–
Manipur	365	–	–
Manipur	166	4	756.42 ha
Meghalaya	79	–	26,326 ha
Orissa	322	–	50 ha
Rajasthan	1	–	83 ha
Rajasthan	9	–	158 ha
Sikkim	56	4	–
Tamil Nadu	10	–	127 ha
Tamil Nadu	3	–	–
Tamil Nadu	1	–	–
Tamil Nadu	448	28	–
Uttar Pradesh	6	–	5500 ha
West Bengal	7	–	2 ha
West Bengal	190	–	15 ha
West Bengal	670	5	–

Table 3: Sacred Groves and their Estimated Area in Meghalaya (After Tiwari *et al.*, 1999, Upadhaya *et al.*, 2005a).

Sl.No.	*Sacred Groves*	*Area (ha)*
	Jaintia Hills	
1.	Blai law	0.01
2.	Poh puja kopatti	4
3.	Poh moorang	20
4.	ka pun Lyngdoh	15
5.	Khlaw Byrsan	50
6.	Poh Lyngdoh	30
7.	Law kyntang	400
8.	khlaw Blai	15
9.	Khloo paiu Ram pyrthai	150
10.	Trepale jowai	70
11.	khloo Lyngdoh	15
12.	Mokhain	45
13.	Dpepati Myndihati	15
14.	Lumtiniangmokaiaw Syndai	25
15.	Lawialong	12
	Ri-Bhoi district	
16.	Pahampdem	900
17.	Nong Lyngdoh	90
18.	Sophetbneng	90
	East Khasi Hills	
19.	Law Lyngdoh, Mawphlong	75
20.	Law Lyngdoh, Smit	6
21.	Lum shyllong	7
22.	Law kyntang	50
23.	Law Adong	90
24.	Khlaw Ram jadong	50
25.	Law Blei Bah	120
26.	Mawlong syiew	120
27.	Pom shandy	80

Contd...

Table 3–Contd...

Sl.No.	Sacred Groves	Area (ha)
28.	Law Adong	400
29.	Law suidnon	80
30.	Law u-Niang	10
31.	Madan Jadu	5
32.	Law Ar- Liang	25
33.	Lum Diengiri	25
34.	Mawmang	15
35.	Wakhem	10
36.	Law Lieng	20
37.	Law mawsaptur	50
38.	Law Dynmiew	200
39.	Law Nongshim	5
40.	Mawsawa	50
41.	Mawryot	40
42.	Niangdoh	30
43.	Rijaw	35
44.	Umtong	400
45.	Diengkain	400
46.	Mawthoh	600
47.	Maw kyrngah	1200
48.	Kynsang	150
49.	Umthri	80
50.	Umkatait	100
	West Khasi Hills	
51.	ULawLyngdoh, Nonglyngkien	90
52.	ULumsanglia, Nonglyngkien	45
53.	ULumBlei, Nonglyngkien	55
54.	Law lyngdoh, Nonglang	200
55.	Law kyntang, Mawlangwir	300
56.	Law kyntang, Mawten	100
57.	Law Lyngdoh, Rangmaw	400

Contd...

Table 3–Contd...

Sl.No.	Sacred Groves	Area (ha)
58.	Lawkyntang, Mawthawiaw	100
59.	Nongsynrih	100
60.	Law Lyngdoh, Nonglait	50
61.	Law Adong, Mawlong	200
62.	Kyllai Lyngun, Mariam	80
63.	Lyngdoh , Mawnai	80
64.	Law Lyngdoh Nongbri	5
65.	Lawkyntang, Lyngdoh lawren	10
66.	Law Lyngdoh Mawlot	20
67.	Law Lyngdoh Wanniang Sawkpoh	7
68.	Dienglienbah, Rngikseh	0.5
69.	Law Lyngdoh Wahlang, Nongklung	10
	East Garo Hills	
70.	Kimpra Hills,Risubakrapara	20
71.	Kondal Hills, Risubakrapara	10
72.	Walchi Ruram Hills, Risubakrapara	25
73.	Ganna Ram–ram rock Megapgiri	30
74.	Miapara Ranggdom	1
75.	Bego Miapara Ranggdom	1
76.	Jongola Miapara	1
77.	Rautagiri	37
	West Garo Hills	
78.	Goragiri	25
79.	Damalgiri	50
80.	Angalgiri	20
81.	Asigiri	4
82.	Jelbongpara	20
83.	Sadoipara	30
84.	Jhanjipara	7
85.	Daronggiri	25

Table 4: Species Composition of the Sacred Groves of Jaintia hills, Meghalaya (The nomenclature of the species recorded in the sacred groves follow regional flora. The names have been arranged in alphabetical order)

Sl.No.	*Name of Species*	*Family*
I.	**Trees (>8m height)**	
1.	*Acer laevigatum* Wall.	Aceraceae
2.	*Acer oblongum* Wall.	Aceraceae
3.	*Actinodaphne obovata* (Nees) Bl.	Lauraceae
4.	*Alangium chinensis* (Lour.) Harms	Coranaceae
5.	*Albizia chinensis* (Osb.) Merr.	Mimosaceae
6.	*Alseodaphne petiolaris* Hk.f.	Lauraceae
7.	*Antidesma bunius* (L.) Spreng	Euphorbiaceae
8.	*Antidesma diandrum* (Roxb) Roth	Euphorbiaceae
9.	*Antidesma khasianum* Hk.f.	Euphorbiaceae
10.	*Beilschmedia assamica* Meissn.	Lauraceae
11.	*Beilschmedia roxburghiana* Nees	Lauraceae
12.	*Betula alnoides* Buch-Ham. ex D.Don.	Betulaceae
13.	*Brassiopsis glomerulata* (Bl.) Regal	Araliaceae
14.	*Calophyllum polyanthum* Choisy	Clusiaceae
15.	*Camellia cauduca* Cl.ex Brandis	Theaceae
16.	*Camellia caudata* Wall.	Theaceae
17.	*Capparis acutifolia* Sweet	Capparaceae
18.	*Caryota urens* Linn.	Araceae
19.	*Castanopsis indica* A.DC.	Fagaceae
20.	*Castanopsis purpurella* (Miq.) Balak	Fagaceae
21.	*Castanopsis tribuloides* (Sm.) DC.	Fagaceae
22.	*Cinnamomum bejolghota* (Buch-Ham.) Sweet.	Lauraceae
23.	*Cinnamomum glanduliflerum* (Wall.) Meissn.	Lauraceae
24.	*Cinnamomum glaucescens* (Nees) Meissn.	Lauraceae
25.	*Cinnamomum tamala* Fr. Nees	Lauraceae
26.	*Citrus latipes* (Swingle) Tanaka	Rutaceae
27.	*Croton oblongus* Burm.f.	Euphorbiaceae
28.	*Cryptocarya floribunda* Nees.	Lauraceae

Contd...

Table 4–Contd...

Sl.No.	Name of Species	Family
29.	*Cythea* sp.	Cytheaceae
30.	*Desmos longiflorus* (Roxb.) Safford	Annonaceae
31.	*Diospyros kaki* Thunb.	Ebenaceae
32.	*Docynia indica* (Wall) Decne	Rosaceae
33.	*Drimycarpus racemosus* (Roxb.) Hk.f.	Anacardiaceae
34.	*Dysoxylum gobara* (Buch-Ham.) Merr.	Meliaceae
35.	*Elaeocarpus lancifolius* Roxb.	Elaeocarpaceae
36.	*Elaeocarpus sikkimensis* Mast.	Elaeocarpaceae
37.	*Engelhardtia spicata* Bl.	Juglandiaceae
38.	*Erythrina arborescens* Roxb.	Fabaceae
39.	*Euonymus attenuatus* Laws.	Celastraceae
40.	*Eurya acuminata* DC.	Theaceae
41.	*Eurya cerasifolia* (D. Don) Kobuski	Theaceae
42.	*Evodia trichitoma* (Lour.) Planch.	Rutaceae
43.	*Ficus altissima* Bl.	Moraceae
44.	*Ficus concinna* Miq.	Moraceae
45.	*Ficus elmeri* Govt.	Moraceae
46.	*Ficus mclelandi* var. *rhododendrifolia* Corn.	Moraceae
47.	*Ficus nerifolia* J.E.Sm	Moraceae
48.	*Ficus virens* Ait.	Moraceae
49.	*Ficus* sp.	Moraceae
50.	*Garcinia Morella* Desr.	Clusiaceae
51.	*Garcinia tinctoria* (DC.) W.F. Wight	Clusiaceae
52.	*Helecia nilagirica* Bedd.	Proteaceae
53.	*Ilex embelioides* Hook.f.	Aqualifoliaceae
54.	*Itea chinensis* Hook.f.	Iteaceae
55.	*Itea macrophylla* Wall.	Iteaceae
56.	*Ixora subsessilis* G.Don.	Rubiaceae
57.	*Knema angustifolia* (Roxb.) Warb.	Myristicaceae
58.	*Ligustrum robustrum* (Roxb.) Bl.	Oleaceae
59.	*Lindera latifolia* Hk.f.	Lauraceae

Contd...

Table 4–Contd...

Sl.No.	Name of Species	Family
60.	*Lindera nacusua* (D.Don.) Merr.	Lauraceae
61.	*Lindera reticulata* Benth.	Lauraceae
62.	*Lithocarpus elagans* (Bl.) Hatus ex Soep.	Fagaceae
63.	*Lithocarpus fenestrate* (Roxb.) Rehder.	Fagaceae
64.	*Lithocarpus* sp.	Fagaceae
65.	*Litsea citrata* Bl.	Lauraceae
66.	*Litsea sebifera* Pers.	Lauraceae
67.	*Litsea semicarpifolia* (Nees) Hook.f.	Lauraceae
68.	*Lyonia ovalifolia* var. *ovalifolia* (Wall.) Drude	Ericaceae
69.	*Macaranga denticulata* Muell.-Arg.	Euphorbiaceae
70.	*Macropanax dispermus* (Bl) O.Ktz.	Araliaceae
71.	*Manglietia insignis* (Wall) Bl.	Magnoliaceae
72.	*Melia azedarach* Linn.	Meliaceae
73.	*Meyna spinosa* Link.	Rubiaceae
74.	*Michelia doltsopa* DC.	Magnoliaceae
75.	*Microtropis discolor* (Wall.) Arn.	Celastraceae
76.	*Myrica esculenta* Buch.-Ham.ex.D.Don	Myricaceae
77.	*Myrsine semiserrata* Wall.	Myrsinaceae
78.	*Neolitsea cassia* (Linn) Koster.	Lauraceae
79.	*Ostodes paniculata* Bl.	Euphorbiaceae
80.	*Persea bombycina* (King ex Hk.f.) Koster	Lauraceae
81.	*Persea duthiei* (King ex Hk.f.) Koster.	Lauraceae
82.	*Persea gamblei* (King ex Hk.f.) Koster.	Lauraceae
83.	*Persea odoratissima* (Nees) Koster.	Lauraceae
84.	*Persea parviflora* (Meissn.) Haridasan et Rao	Lauraceae
85.	*Phoebe lanceolata* (Nees)	Lauraceae
86.	*Picresema* sp.	Simaroubiaceae
87.	*Pinus kesiya* Royle.ex Gordon.	Pinaceae
88.	*Pithecellobium monadelphum* (Roxb.) Koster.	Mimosaceae
89.	*Pittosporum podocarpum* Gagnep.	Pittosporaceae
90.	*Podocarpus nerifolia* D. Don.	Podocarpaceae

Contd...

Table 4–Contd...

Sl.No.	Name of Species	Family
91.	*Prunus acuminata* (Wall.) Dietr.	Rosaceae
92.	*Prunus jenkinsii* Hook.f.	Rosaceae
93.	*Pseudostreblus indica* Bureau.	Moraceae
94.	*Pyrularia edulis* A. DC.	Santalaceae
95.	*Pyrus pashia* D.Don.	Rosaceae
96.	*Quercus griffithii* Hk.f.& Th ex DC.	Fagaceae
97.	*Quercus serrata* Thunb.	Fagaceae
98.	*Randia griffithii* Hk.f.	Rubiaceae
99.	*Rhododendron arboreum* Sm.	Ericaceae
100.	*Rhus acuminata* DC.	Anacardiaceae
101.	*Rhus hookeri* Sahna and Bahad.	Anacardiaceae
102.	*Sapindus rarak* DC.	Sapindaceae
103.	*Sarcosperma griffithii* Clarke	Sapotaceae
104.	*Schefflera elata* (Buch.-Ham.) Harms	Araliaceae
105.	*Schefflera hypoleuca* (Kurz) Harms	Araliaceae
106.	*Schima khasiana* Dyer.	Theaceae
107.	*Schima wallichii* (DC.) Korth	Theaceae
108.	*Spondias axallaris* Roxb.	Anacardiaceae
109.	*Sterculia hamiltonii* (O.Ktze.) Adelb	Sterculiaceae
110.	*Styrax hookerii* Cl	Styracaceae
111.	*Styrax serrulatum* Roxb.	Styracaceae
112.	*Symplocos pyrifolia* Wall.ex G.Don	Symplocaceae
113.	*Syzygium tetragonum* (Wt.) Kurz	Myrtaceae
114.	*Vaccinium sperengelii* (G.Don) Rehd.	Vaccinaceae
115.	*Viburnum simonsii* Hk.f.& Th.	Caprifoliaceae
116.	*Wendlendia wallichii* W.& A.Prodr.	Rubiaceae
117.	*Zanthoxylum acanthopodium* DC.	Rutaceae
118.	*Castanopsis kurzii* (Hance) S.N. Biswas	Fagaceae
119.	*Elaeocarpus floribundus* Bl.	Elaeocarpaceae
120.	*Ficus oliogodon* Miq.	Moraceae
121.	*Litsea laeta* (Nees) Hk.f.	Lauraceae

Contd...

Table 4–Contd...

Sl.No.	Name of Species	Family
122.	*Neolitsea cuipala* (D. Don) Koster	Lauraceae
123.	*Persea globularia* Koster	Lauraceae
124.	*Prunus nepalensis* (Ser.) Steud.	Rosaceae
125.	*Quercus glauca* Thunb.	Fagaceae
126.	*Ficus lamponga* Miq.	Moraceae
127.	*Garcinia anomala* Planch.& Triana	Clusiaceae
128.	*Ilex venulosa* Hk.f.	Aqualifoliaceae
129.	*Lithocarpus dealbatus* (Miq) Rehder.	Fagaceae
130.	*Premna racemosa* Wall ex Sch. scandens Roxb.	Verbenaceae
131.	*Quercus kamroopii* D.Don	Fagaceae
132.	*Symplocos paniculata* (Thunb)Miq.	Symplocaceae
133.	*Symplocos racemosa* Roxb.	Symplocaceae
134.	*Symplocos viridissima* Brandis	Symplocaceae
135.	*Fraxinus floribunda* Wall.	Oleaceae
136.	*Ternstroemia gymnanthera* (W.&A.)Bedd.	Theaceae
137.	*Xerospermum gymnanthera*	Theaceae
138.	*Acrmychia pedunculata* (L.) Miq.	Rutaceae
139.	*Carpinus viminea* Lindl.	Corylaceae
140.	*Glochidion lanceolarium* (Roxb)Voight	Euphorbiaceae
141.	*Glochidion thomsonii* (Muell-Arg) Hk.f.	Euphorbiaceae
142.	*Pentapanax subcordatus* (G.Don) Seem.	Araliaceae
143.	*Randia wallichii* Hk.f.	Rubiaceae
144.	*Xantolis assamica* (Clarke) Van Royan.	Sapotaceae
II.	**Small trees or Large shrubs (<8m)**	
1.	*Aralia thomsonii* Seem	Araliaceae
2.	*Cinnamomum pauciflorum* Nees	Lauraceae
3.	*Clerodendrum bracteatum* Wall.ex Walp	Verbenaceae
4.	*Coffea khasiana* Hook.f.	Rubiaceae
5.	*Corylopsis himalayana* Griff.	Hamamelidaceae
6.	*Cythea* sp.	Cytheaceae
7.	*Erythroxylum kunthianum* Wall.ex Kurz.	Erythroxylaceae

Contd...

Table 4–Contd...

Sl.No.	Name of Species	Family
8.	*Euonymus lowsonii* Clarke and Prain	Celastraceae
9.	*Eurya japonica* Thunb.	Theaceae
10.	*Ficus hirta* var. *roxburghii* (Miq.) King	Moraceae
11.	*Leucosceptrum canum* Smith	Lamiaceae
12.	*Lindera caudata* Benth.	Lauraceae
13.	*Litsea salicifolia* (Roxb.ex Nees) Hk.f.	Lauraceae
14.	*Maesa indica* (Roxb.) Wall.	Myrsinaceae
15.	*Micromelum integerrifolia* (Roxb.) Roem.	Rutaceae
16.	*Paramignya micrantha* Kurz.	Rutaceae
17.	*Pouzolzia frondosa* (D.Don) O.Ktze.	Urticaceae
18.	*Pseudobrassaiopsis hispida* (Seem) R.N.Ban.	Araliaceae
19.	*Rhus javanica* Linn.	Anacardiaceae
20.	*Saurauia punduana* Wall.	Saurauiaceae
21.	*Symplocos crataegoides* D.Don	Symplocaceae
22.	*Symplocos spicata* Roxb.	Symplocaceae
23.	*Viburnum foetidum* Wall.	Caprifoliaceae
24.	*Aralia armata* (G.Don) Seem	Araliaceae
25.	*Ardisia depressa* Cl.	Myrsinaceae
26.	*Camellia kissi* Wall.	Theaceae
27.	*Clerodendron calebrookienum* Walp.	Verbenaceae
28.	*Diospyrus motana*	Ebenaceae
29.	*Pittosporum reticosum* Redl.	Pittosporaceae
30.	*Syzygium macrocarpum* (Roxb.) Balakr.	Myrtaceae
31.	*Bauhinia variegata*	Caesalpinaceae
32.	*Daphne composite* (L.) Gilg.	Thymeliaceae
33.	*Debregeasia longifolia* (Brum.f.) Wedl.	Urticaceae
34.	*Merrilliopanax japonicum* Seem.	Araliaceae
35.	*Pandanus odoratissimus* L.f.	Pandanaceae
36.	*Polygala arillata* Buch-Ham.	Polygalaceae
37.	*Sambucus javanica* Bl. Bijdr	Caprifoliaceae
38.	*Turpina nepalensis* W and A Prodr.	Staphylaceae

Contd...

Table 4–Contd...

Sl.No.	Name of Species	Family
III.	**Shrubs**	
1.	*Achyranthes aspera* L.	Amaranthaceae
2.	*Ardisia griffithii* Cl.	Myrsinaceae
3.	*Artemesia nilagirica* (Clarke) Pamp.	Asteraceae
4.	*Boehmeria platyphylla* D.Don.	Urticaceae
5.	*Boehmeria sidaefolia* Wedd.	Urticaceae
6.	*Breynia retusa* (Dennst) Alst.	Euphorbiaceae
7.	*Brucea mollis* Wall.ex Kurz.	Sibaroubiaceae
8.	*Callicarpa psilocalyx* Clarke	Verbenaceae
9.	*Callicarpa rubella* Lindl.	Verbenaceae
10.	*Cassia floribunda* Cav.	Fabaceae
11.	*Clerodendrum lastocephalum* Clarke	Verbenaceae
12.	*Clerodendrum wallichii* Merr.	Verbenaceae
13.	*Crotolaria anagyroides* HBK.	Fabaceae
14.	*Datura stromonium* Linn.	Solanaceae
15.	*Desmodium floribundum* (D.Don) G. Don	Fabaceae
16.	*Desmodium heterocarpum* (Linn.) DC.	Fabaceae
17.	*Eupatorium adenophorum* Spreng.	Asteraceae
18.	*Ficus clavata* Wall ex Miq.	Moraceae
19.	*Ficus fulva* Reinwdt	Moraceae
20.	*Ficus gasperriniana* Miq.	Moraceae
21.	*Flemingia macrophylla* (Willd.) Prain	Fabaceae
22.	*Girardinia palmata* (Forsk) Gaud.	Urticaceae
23.	*Goldfussia glabrata* (Nees) Balakr.	Acanthaceae
24.	*Gomphostemma parviflorum* Wall.ex Benth.	Lamiaceae
25.	*Goniothalamus sesquipedalis* (Wall.) Hk.f.& Th.	Annonaceae
26.	*Inula cappa* (D.Don) DC. Prodr.	Asteraceae
27.	*Lantana camara* Linn.	Verbenaceae
28.	*Lasianthus lucidus* Bl.	Rubiaceae
29.	*Lasianthus sikkimensis* Hook.f.	Rubiaceae
30.	*Mahonia pycnophylla* (Fedde) Takeda	Berberidaceae

Contd...

Table 4–Contd...

Sl.No.	Name of Species	Family
31.	*Melastoma nepalensis* Lodd.	Melastomaceae
32.	*Munronia pinnata* (Wall.) Harms.	Meliaceae
33.	*Mussaenda roxburghii* Hk.f.	Rubiaceae
34.	*Neillia thyrsiflora* D.Don	Rosaceae
35.	*Phlogacanthus pubinervius* T.Anders.	Acanthaceae
36.	*Phyllanthus* sp.	Euphorbiaceae
37.	*Plectranthus striatus* Benth.	Lamiaceae
38.	*Prinsepia utilis* Royle	Rosaceae
39.	*Pseuderanthemum indicum* A.M.&J.M.Cowan	Acanthaceae
40.	*Psychotria curviflora* Wall.	Rubiaceae
41.	*Psychotria erratica* Hook.f.	Rubiaceae
42.	*Rhynchotecum vestitum* Wall.ex Cl.	Gesneriaceae
43.	*Rhynchotecum ellipticum* (Wall ex Dietr) A.DC	Gesneriaceae
44.	*Rauvolfia densiflora* (Wall.) Benth.ex Hk.f.	Apocynaceae
45.	*Sarcandra glabra* (Thunb.) Nakai	Chloranthaceae
46.	*Senecio cappa* Buch-Ham.ex D.Don	Asteraceae
47.	*Sida rhombifolia* Linn.	Malvaceae
48.	*Smilax myrtillus* DC.	Smilacaceae
49.	*Solanum aculeatissimum* Jacq.	Solanceae
50.	*Sympagia monadelpha* (Nees) Bremek.	Acanthaceae
51.	*Teucrium quadrifarium* Buch-Ham.	Lamiaceae
52.	*Triumfetta tomentosa* Bojer	Tiliaceae
53.	*Urena lobata* Linn.	Malvaceae
54.	*Blumea balsamifera* (L) DC.	Asteraceae
55.	*Blumea procera* DC.	Asteraceae
56.	*Buddleja asiatica* Lour.	Buddljeaceae
57.	*Chimonobambusa khasiana* (Munro.) Nakai	Poaceae
58.	*Daphne involucrata* Wall.	Thymeliaceae
59.	*Ficus prostrate*	Moraceae
60.	*Hydrangea vestita* Wall.	Hydrangeaceae
61.	*Inula eupatoriades* DC. Prodr.	Asteraceae

Contd...

Table 4–Contd...

Sl.No.	Name of Species	Family
62.	*Lobelia nicotianifolia* R&S Syst	Lobeliaceae
63.	*Luculia pinceana* Hk.	Rubiaceae
64.	*Maesa tetrandra* (Roxb.) DC.Prodr.	Myrsinaceae
65.	*Priotropis cytisoides* (Roxb.) W&A Prodr.	Fabaceae
66.	*Prismatomeris tetrandra* (Roxb.) K.Schun	Rubiaceae
67.	*Senecio jowaiensis* Balakr.	Asteraceae
68.	*Themeda arundinaceae* (Roxb.)Ridley	Poaceae
69.	*Thysanolaena maxima* (Roxb.) O. Ktze.	Poaceae
70.	*Vernonia volkameriifolia* DC.Prodr.	Asteraceae
71.	*Artemesia parviflora* Roxb.	Asteraceae
72.	*Baliospermum micranthum* Muell-Arg.	Euphorbiaceae
73.	*Desmodium gyriodes* (Link) DC.	Fabaceae
74.	*Desmodium microphyllum* (Thunb) DC.Prodr.	Fabaceae
75	*Psychotria monticola* Kurz.	Rubiaceae
76.	*Psychotria symplocifolia* Kurz.	Rubiaceae
IV.	**Scandent shrub**	
1.	*Alyxia fascicularis* Benth.	Apocynaceae
2.	*Capparis* sp.	Capparaceae
3.	*Celastrus stylosus* Wall.	Cellastraceae
4.	*Embelia subcoriaceae* (Clarke) Mez.	Myrsinaceae
5.	*Eleagnus pyriformis* Hook.f.	Eleagnaceae
6.	*Embelia parviflora* DC.	Myrsinaceae
7.	*Embelia vestita* Roxb.	Myrsinaceae
8.	*Euonymus theifolius* Wall. ex Laws.	Celestraceae
9.	*Jasminium dispermum* Wall.	Oleaceae
10.	*Jasminum lanceolaria* Roxb.	Oleaceae
11.	*Jasminum nepalense* Spreng.	Oleaceae
12.	*Lygodium flexuosum* (Linn.) Sw.	Lygodiaceae
13.	*Polygonum molle* D.Don	Polygonaceae
14.	*Polygonum perfoliatum* Linn.	Polygonaceae
15.	*Rubus alceifolius* Poir.	Rosaceae

Contd...

Table 4–Contd...

Sl.No.	*Name of Species*	*Family*
16.	*Rubus assamensis* Focke	Rosaceae
17.	*Rubus ellipticus* Smith	Rosaceae
18.	*Rubus khasianus* Cordat.	Rosaceae
19.	*Smilax aspera* L.	Smilacaceae
20.	*Smilax lanceiofolia* Roxb.	Smilacaceae
21.	*Smilax perfoliata* Lour.	Smilacaceae
22.	*Toddalia asiatica* (Linn.) Lamk.	Rutaceae
23.	*Tupidanthus calyptratus* Hk.f.& Th.	Araliaceae
24.	*Zanthoxylum oxyphyllum* Edgew.	Rutaceae
25.	*Blumea riparia* (Bl.) DC.	Asteraceae
26.	*Ficus sarmentosa* Ham.ex.Sm.	Moraceae
27.	*Morinda umbelleta* L.Sp.	Rubiaceae
28.	*Rubus niveus* Thunb.	Rosaceae
29.	*Rubus rugosus* Smith.	Rosaceae
30.	*Alyxia graciles* (DC) Hk.f.	Apocynaceae
V.	**Climbers**	
1.	*Acacia pennata* (Linn.) Willd.	Mimosaceae
2.	*Cayratia japonica* (Thunb.) Gagnep.	Vitaceae
3.	*Cissampelos pareire* Linn.	Menispermaceae
4.	*Cissus repens* Lamk.	Vitaceae
5.	*Clematis buchaniana* DC.	Ranunculaceae
6.	*Cyclea bicristata* (Griff.) Diels.	Menispermaceae
7.	*Dioscorea bulbifera* Linn.	Dioscoreaceae
8.	*Dioscorea kamoonesis* Kunth.	Dioscoreaceae
9.	*Entada phaseoloides* (Linn.) Merr.	Mimosaceae
10.	*Fissistigma verrucosum* (Hk.f.& Th) Merr.	Annonaceae
11.	*Gnetum montanum* Mg.F.	Gnetaceae
12.	*Hedera nepalensis* K.Koch.	Araliaceae
13.	*Hedyotis scandens* D.Don	Rubiaceae
14.	*Heterostemma alatum* Wt.	Asclepiadiaceae
15.	*Hodgsonia macrocarpa* (Bl.) Cogn.	Cucurbitaceae

Contd...

Table 4–Contd...

Sl.No.	*Name of Species*	*Family*
16.	*Holboellia latifolia* Wall.	Lardizabalaceae
17.	*Melodinus khasianus* Hk.f.	Apocynaceae
18.	*Melodinus monogynus* Roxb.	Apocynaceae
19.	*Millettia cinerea* Benth.	Fabaceae
20.	*Paederia foetida* Linn.	Rubiaceae
21.	*Passiflora edultes* Sims.	Passifloraceae
22.	*Passiflora nepalensis* Wall.	Passifloraceae
23.	*Passiflora stipulate* Aubl.	Passifloraceae
24.	*Pericampylus glaucus* (Lamk.) Merr.	Menispermaceae
25.	*Piper griffithii* C.DC.	Piperaceae
26.	*Piper mullesua* D.Don	Piperaceae
27.	*Piper peepuloides* Roxb.	Piperaceae
28.	*Rourea minor* (Gaert.) Leenh.	Connaraceae
29.	*Rubia cordifolia* Linn.	Rubiaceae
30.	*Shuteria vestita* W.& A.	Fabaceae
31.	*Stephania glandulifera* Miers.	Menispermaceae
32.	*Tetrastigma leucostaphyllum* (Dennst) Balak.	Vitaceae
33.	*Tetrastigma serrulatum* (Roxb.) Planch.	Vitaceae
34.	*Trichosanthes bracteata* (Lamk.) Voigt.	Cucurbitaceae
35.	*Trichosanthes wallichiana* (Ser) Wt.	Cucurbitaceae
36.	*Trichosanthes himalensis* Clarke	Cucurbitaceae
37.	*Vitis heynenna* Roem. and Schult	Vitaceae
38.	*Actinidia callosa* Lindl.	Actinidiaceae
39.	*Ceropegia angustifolia* Wt.	Asclepediaceae
40.	*Clematis acutangula* Hk.f.	Ranunculaceae
41.	*Clematis acuminata* DC. Syst.	Ranunculaceae
42.	*Edysanthera micrantha* DC. Prodr.	Apocynaceae
43.	*Ipomea nil* (Linn.) Roth.	Convolvulaceae
44.	*Lonicera maciantha* DC.	Caprifoliaceae
45.	*Vallaris solanacea* (Roth) O. Ktze	Apocynaceae

Contd...

Table 4–Contd...

Sl.No.	*Name of Species*	*Family*
VI.	**Epiphytes**	
1.	*Aerides multiflora* Roxb.	Orchidaceae
2.	*Aeschynanthes parasiticus* (Roxb.) Wall	Gesneriaceae
3.	*Aeschynanthes sikkimensis* (Clarke) Stapf.	Gesneriaceae
4.	*Aeschynanthes superba* Clarke	Gesneriaceae
5.	*Agapetes auriculata* Hk.f.	Vaccinaceae
6.	*Agapetes variegata* (Roxb.) G.Don.	Vaccinaceae
7.	*Agrostaphyllum brevipes* King and Pantl.	Orchidaceae
8.	*Asplenium phyllitides* D.Don	Aspleniaceae
9.	*Bulbophyllum affine* Lindl.	Orchidaceae
10.	*Bulbophyllum griffithii* (Lindl.) Reichb.f.	Orchidaceae
11.	*Bulbophyllum gymnopus* Hk.f.	Orchidaceae
12.	*Coelogyne stricta* (Don) Schltr.	Orchidaceae
13.	*Coelogyne suaveolens* (Lindl.) Hk.f.	Orchidaceae
14.	*Cryptochilus sanguinea* Wall.	Orchidaceae
15.	*Crypsinus hastata* (Thunb) Copel.	Polypodiaceae
16.	*Cymbidium* sp.	Orchidaceae
17.	*Dendrobium chrysanthum* Lindl.	Orchidaceae
18.	*Dendrobium devonianum* Paxt.	Orchidaceae
19.	*Dendrobium nobile* Lindl.	Orchidaceae
20.	*Dendrobium ochreaceum* Lindl.	Orchidaceae
21.	*Dendrobium williamsonii* Day and Reichb. F.	Orchidaceae
22.	*Euonymus frigidus* Wall.	Celastraceae
23.	*Helixanthera ligustrina* (Wall.) Danser	Loranthaceae
24.	*Hoya lanceolata* Wall.	Asclepiadiaceae
25.	*Hoya purviflora* Wight.	Asclepiadiaceae
26.	*Hoya* sp.	Asclepiadiaceae
27.	*Lepidogrammitis rostrata* (Hook.) Ching	Polypodiaceae
28.	*Loranthus umbellifer* Schult.	Loranthaceae
29.	*Luisia inconspicua* (Hk.f.) King and Prantl.	Orchidaceae
30.	*Luisia psyche* Reichb.	Orchidaceae

Contd...

Table 4–Contd...

Sl.No.	Name of Species	Family
31.	*Lycopodium setaceum* Buch.-Ham.	Lycopodiaceae
32.	*Lysionotus serratus* D.Don.	Gesneriaceae
33.	*Microsorum superficiale* (D.Don) Ching	Polypodiaceae
34.	*Oberonia tridifolia*	Orchidaceae
35.	*Pholiodata imbricata* Hook.	Orchidaceae
36.	*Pholiodata undulate* Lindl.	Orchidaceae
37.	*Pleione maculata* (Smith) D.Don.	Orchidaceae
38.	*Pleione praecox* (Smith) D.Don.	Orchidaceae
39.	*Pothos scandens* Linn.	Araceae
40.	*Pyrrosia heteractis* (Mett ex Kuhn) Ching	Polypodiaceae
41.	*Rhaphidophora calophyllum* Schott.	Araceae
42.	*Rhaphidophora decursiva* (Roxb.) Schott.	Araceae
43.	*Schefflera venulosa* (W. and A.) Harms.	Araliaceae
44.	*Tainia viridifusca* (Hook) Benth.	Orchidaceae
45.	*Vaccinium vacciniaceum* (Roxb.) Sleum.	Ericaceae
46.	*Viscum monoicum* Roxb. ex DC.	Loranthaceae
47.	*Bulbophyllum cylindraceum* Lindl.	Orchidaceae
48.	*Bulbophyllum reptans* Lindl.	Orchidaceae
49.	*Coelogyne viscose* Reichb.	Orchidaceae
50.	*Cymbidium cochleare* Lindl.	Orchidaceae
51.	*Cymbidium lancifolium* Hk.	Orchidaceae
52.	*Dendrobium acinaciforme* (Roxb.)	Orchidaceae
53.	*Dendrobium pauciflorum* King and Prantl.	Orchidaceae
54.	*Dendrobium williamsonii* Day& Reid.	Orchidaceae
55.	*Eria bambusifolia* Lindl.	Orchidaceae
56.	*Macrosolen psilanthus* (Hook.f.) Danse	Loranthaceae
57.	*Otochilus alba* Lindl.	Orchidaceae
58.	*Otochilus fusca* Lindl.	Orchidaceae
VII.	**Herbs**	
1.	*Aeginetia indica* Linn.	Orobanchaceae
2.	*Achyrospermum wallichianum* (Benth.) Hk.f.	Lamiaceae
3.	*Ageratum conyzoides* Linn.	Asteraceae

Contd...

Table 4–Contd...

Sl.No.	Name of Species	Family
4.	*Ainsliaea latifolia* (D. Don) Sch.	Asteraceae
5.	*Anaphalis adnata* DC.	Asteraceae
6.	*Anaphalis contorta* (D. Don) Hk.f.	Asteraceae
7.	*Anemone rivularis* Ham.	Ranunculaceae
8.	*Anoectochilous roxburghii* (Wall.) Lindl.	Orchidaceae
9.	*Anotis wightiana* Hk.f.	Rubiaceae
10.	*Arisaema consanguineum* Schott.	Araceae
11.	*Arisaema tortusum* (Wall.) Schott.	Araceae
12.	*Athyriopsis japonica* (Thunb.) Ching	Athyriaceae
13.	*Arundinella nepalensis* Trin.	Poaceae
14.	*Asparagus filicinus* Buch.-Ham.	Liliaceae
15.	*Asplenium cheilosorum* Kuntze	Aspleniaceae
16.	*Asplenium unilaterale* Lamk.	Aspleniaceae
17.	*Athyrium drepanopterum* (Kunize) A. Brown	Athyriaceae
18.	*Balanophora dioca* R.Br.	Balanophoraceae
19.	*Begonia palmate* D.Don	Begoniaceae
20.	*Begonia* sp.	Begoniaceae
21.	*Bidens pilosa* (Bl.) Sherff	Asteraceae
22.	*Blumea hieraciifolia* (D.Don) DC.	Asteraceae
23.	*Borreria articularis* (L.f) F.N. Williams	Rubiaceae
24.	*Calanthe clavata* Lindl.	Orchidaceae
25.	*Cardamine hirsuta* Linn.	Brassicaceae
26.	*Carex filicina* Nees.	Cyperaceae
27.	*Centella asiatica* (Linn.) Urban	Apiaceae
28.	*Chirita oblongifolia* (Roxb.) Sinclair	Gesneriaceae
29.	*Cirsium involucratum* DC.	Asteraceae
30.	*Colosasia esculenta* (Linn.) Schott	Araceae
31.	*Commelina paludosca* Bl.	Commelinaceae
32.	*Costus speciosus* (Keonig.) Smith	Zingiberaceae
33.	*Crassocephalum crepidioides* (Benth.) Moore	Asteraceae
34.	*Cyanotis vaga* (Lour.) J.A. and J.H.Schult.	Commelinaceae

Contd...

Table 4–Contd...

Sl.No.	Name of Species	Family
35.	*Cyperus flavidus* Retz.	Cyperaceae
36.	*Cyperus kyllingia* Endl.	Cyperaceae
37.	*Cyperus pilosus* Vahl	Cyperaceae
38.	*Dianella ensata* (Thunb.) R.J. Handerson	Liliaceae
39.	*Dichanthium coricosum* (L.) A. Camus	Poaceae
40.	*Dicranopteris linearis* (Burm.f) Underwood	Gleicheniaceae
41.	*Dicrocephala bicolor* (Roth) Schltdl.	Asteraceae
42.	*Digitaria ciliaris* (Retz.) Koel.	Poaceae
43.	*Dipteris wallichii* (R.Br.) Moore	Dipteridaceae
44.	*Disporum cantoniense* (Lour.) Merr.	Liliaceae
45.	*Drosera peltata* Smith.	Droceraceae
46.	*Drymaria cordata* (Linn.) Roem and Schult.	Caryophyllaceae
47.	*Elatostema dissectum* Wedd.	Urticaceae
48.	*Elatostema hookerianum* Wedd.	Urticaceae
49.	*Elatostema sessile* Frost.	Urticaceae
50.	*Elsholtzia blanda* (Benth.) Benth	Lamiaceae
51.	*Elsholtzia strobilifera* (Benth.) Benth	Lamiaceae
52.	*Emilia sonchifolia* (Linn.) DC.	Asteraceae
53.	*Eragrostis gangetica* (Roxb.) Steud.	Poaceae
54.	*Eragrostis unioloides* (Retz.) Steud.	Poaceae
55.	*Eriosaema himalaicum* Ohashi	Fabaceae
56.	*Equisetum ramosissimum* Desf.	Equisetaceae
57.	*Floscopa scandens* Lour.	Commelinaceae
58.	*Forrestia mollissima* (Bl.) Kds.	Commelinaceae
59.	*Galinsoga parviflora* Cav.	Asteraceae
60.	*Geranium nepalense* Sweet.	Geraniaceae
61.	*Gerbera piloselloides* (linn.) Cass.	Asteraceae
62.	*Globba clarkii* Baker.	Zingiberaceae
63.	*Gnaphalium pensylvanicum* Willd.	Asteraceae
64.	*Gomphostemma nutans* Hook.f.	Lamiaceae
65.	*Gomphostemma velutinum* Benth.	Lamiaceae

Contd...

Table 4–Contd...

Sl.No.	Name of Species	Family
66.	*Gynura cusimbusa* (D.Don) S. Moore	Asteraceae
67.	*Habenaria stenostachya* (Lindl.) Benth.	Orchidaceae
68.	Hedychium coccineum Smith.	Zingiberaceae
69.	*Hedychium coronarium* Koen	Zingiberaceae
70.	*Hedychium ellepticum* Smith.	Zingiberaceae
71.	*Hedychium stenopetalum* Lodd.	Zingiberaceae
72.	*Hedyotis tenelliflora* Bl.	Rubiaceae
73.	*Hedyotis uncinella* Hook. and Arn.	Rubiaceae
74.	*Hydrocotyl javanica* Thunb.	Apiaceae
75.	*Hypericum elodeoides* Choisy	Hypericaceae
76.	*Impatiens benthamii* V. Steenis	Balsaminaceae
77.	*Impatiens chinensis* Linn.	Balsaminaceae
78.	*Impatiens drepanophora* Hook.f.	Balsaminaceae
79.	*Impatiens juripa* Hk.f. and Th.	Balsaminaceae
80.	*Impatiens khasiana* Hk.f.	Balsaminaceae
81.	*Impatiens laevigata* Hook.f. and Th.	Balsaminaceae
82.	*Imperata cylindrica* (Linn.) P. Beanv.	Poaceae
83.	*Isachne himalaica* Hook.f.	Poaceae
84.	*Kaulina pteropus* (Blume) Nayar	Polypodiaceae
85.	*Leucas ciliata* Benth.	Lamiaceae
86.	*Linderbergia muraria* (Roxb.) Bruhl	Schrophulariaceae
87.	*Lobelia angulata* Forst.	Campanulaceae
88.	*Lophatherum gracile* Brogn.	Poaceae
89.	*Lycopodium cernum* Linn.	Lycopodiaceae
90.	*Malaxis latifolia* Sm.	Orchidaceae
91.	*Microsorum membranaceum* (D.Don) Ching	Polypodiaceae
92.	*Murdannia gigantea* (Vahl.) Bruck.	Commelinaceae
93.	*Murdannia nudiflora* (Linn.) Brenan	Commelinaceae
94.	*Neocheiropteris normalis* (D.Don) Togawa	Polypodiaceae
95.	*Ophiopogon parviflorus* (Hk.f.) Hara	Liliaceae
96.	*Ophiorrhiza mungos* L.	Rubiaceae

Contd...

Table 4–Contd...

Sl.No.	Name of Species	Family
97.	*Ophiorrhiza nutans* Clarke	Rubiaceae
98.	*Osbeckia capitata* Benth.	Melastomaceae
99.	*Oxalis corniculata* L.	Oxalidaceae
100.	*Panicum brevifolium* L.	Poaceae
101.	*Peliosanthes bakerii* Hk.f.	Liliaceae
102.	*Peperomia reflexa* A. Dietr.	Piperaceae
103.	*Phrynium placentarium* (Lour.) Merr.	Maranthaceae
104.	*Phyllamthus urinaria* L.	Euphorbiaceae
105.	*Phytolacca acinosa* Roxb.	Phytolaccaceae
106.	*Pilea umbrosa* Wedd.	Urticaceae
107.	*Plantago erosa* Wall.	Plantinaceae
108.	*Pogostemon strig* (Benth.) Benth	Lamiaceae
109.	*Polygonum barbata* L.	Polygonaceae
110.	*Polygonum capitatum* D.Don.	Polygonaceae
111.	*Polygonum dibotyrs* D.Don	Polygonaceae
112.	*Polygonum viscosum* D.Don	Polygonaceae
113.	*Polystichum aculeatum* (Linn.) Roth	Aspidiaceae
114.	*Polyura geminate* Hook.f.	Rubiaceae
115.	*Potentilla fulgens* Wall.	Rosaceae
116.	*Pouzolzia hirta* (Bl.) Hassk.	Urticaceae
117.	*Pronephrium nudatam* (Roxb.) Holttum	Thelypteridaceae
118.	*Pteris quadriaurita* Retz.	Pteridaceae
119.	*Ranunculus cantoniensis* DC.	Ranunculaceae
120.	*Scutelleria discolor* Benth.	Lamiaceae
121.	*Selginella semicordata* (Wall ex. Hk. Et . Grev.) Spring	Selaginellaceae
122.	*Selginella willdenovii* (Desu.) Beker	Selaginellaceae
123.	*Smithia ciliata* Royle	Fabaceae
124.	*Solanum myriacanthum* Dunal.	Solanaceae
125.	*Sonerila khasiana* Clarke	Melastomaceae
126.	*Spilanthes paniculata* DC.	Asteraceae
127.	*Spiranthes sinense* (Pers.) Ames	Orchidaceae

Contd...

Table 4–Contd...

Sl.No.	Name of Species	Family
128.	*Strobalanthes* sp.	Acanthaceae
129.	*Torenia diffusa* D.Don	Schrophulariaceae
130.	*Viola palmaris* Ging.	Violaceae
131.	*Viola sikkimensis* W. Becker	Violaceae
132.	*Zingiber rubens* Roxb.	Zingiberaceae
133.	*Arundinaria subereeta* Munro.	Poaceae
134.	*Carex cruciata* Wahlenb.	Cyperaceae
135.	*Cassia mimosoides* L.	Caesalpinaceae
136.	*Dracaena spicata* Roxb.	Agavaceae
137.	*Elshaltzia stachyodia* (Link) Raiz and Sax.	Lamiaceae
138.	*Eusteralis linaris* (Benth) Panigrahi.	Lamiaceae
139.	*Gerbera maxima* (D.Don) Beauv.	Asteraceae
140.	*Hedychium gardnerianum* Rosc.	Zingiberaceae
141.	*Impatiens acuminata* Hook.f &Th.	Balsaminaceae
142.	*Impatiens porrecta* Hk,f.	Balsaminaceae
143.	*Impatiens pulchra* Hk.f and Th.	Balsaminaceae
144.	*Molineria capitulate* (Lour) Herb.	Hypoxidaceae
145.	*Nicandra physaloides* (L.) Gaertn.	Solanaceae
146.	*Osbeckia stellata* Ker Grawl	Melastomaceae
147.	*Phyllanthus roeperians* Mull-Arg.	Euphorbiaceae
148.	*Polygala sibirica* L.	Polygalaceae
149.	*Polygonum posumbu* D.Don	Polygonaceae
150.	*Salomonia cantoniensis* Lour.	Polygalaceae
151.	*Senecia latiligulatus* Balakr.	Asteraceae
152.	*Vernonia cinerea* (L.) Nees.	Asteraceae
153.	*Pyrrosia rummularifolius* (Sw.) Ching	Polypodiaceae
154.	*Monochasorium subdigitatum* (Blume) Kuhn.	Dennstaedtiaceae
155.	*Pronepherium nudatum* (Roxb.) Helttum	Thelypteridaceae
156.	*Diacaepa aspidioides* Blume	Athyriaceae

Table 5: Floristic Composition of Sohrarim and Swer Sacred Groves in East Khasi Hills, Meghalaya

Sl.No.	Name of Species	Family	Sohrarim	Swer
I.	**Trees**			
1.	*Acer laevigatum* Wall.	Aceraceae	+	–
2.	*Albizia odoratissima* Benth.	Fabaceae	+	–
3.	*Beilschmidia assamica* Meissn.	Lauraceae	+	–
4.	*Brassaiopsis speciosa* Dene and Planch.	Araliaceae	+	–
5.	*Camellia caduca* C.B. Clarke.	Theaceae	+	+
6.	*Camellia drupifera* Lour.	Theaceae	+	+
7.	*Corylopsis himalayana* Griff.	Hamamelidaceae	+	–
8.	*Castanopsis purpurella* (Miq) Balak.	Fagaceae	+	–
9.	*Castanopsis tribuloides* A. DC.	Fagaceae	+	–
10.	*Cinnamomum bejolghota* . (Buch.-Ham) Sweet	Lauraceae	–	+
11.	*Cinnamomum glanduliferum* (Wall) Meissn.	Lauraceae	+	+
12.	*Cinnamomum pauciflorum* Nees.	Lauraceae	+	–
13.	*Cinnamomum zeylanicum* Blume.	Lauraceae	+	–
14.	*Daphne involucrata* Wall.	Thymeliaceae	+	+
15.	*Daphne papyracea* Wall ex Steud.	Thymeliaceae	+	+
16.	*Daphniphyllum himalayense* Muell.	Daphniphyllaceae	+	+
17.	*Elaeocarpus acuminatus* Wall.	Elaeocarpaceae	+	–
18.	*Elaeocarpus lanceaefolious* Roxb.	Elaeocarpaceae	+	+
19.	*Engelhardtia spicata* Leschn.ex Bl.	Juglandaceae	–	+
20.	*Erythroxylon kunthianum* Wall.	Erythroxylaceae	+	+
21.	*Eurya acuminata* DC.	Theaceae	+	+
22.	*Exbucklandia populnea* Br.	Hamamelidaceae	+	–
23.	*Ficus nerifolia* J.E Sm.	Moraceae	+	+
24.	*Garcinia cowa* Roxb.	Clusiaceae	+	–
25.	*Glochidion accuminatum* Muell.Arg.	Euphorbiaceae	+	+
26.	*Glochidion assamicum* Hk.f.	Euphorbiaceae	–	+
27.	*Helicia nilagirica* Bedd.	Proteaceae	+	–

Contd...

Table 5–Contd...

Sl.No.	Name of Species	Family	Sohrarim	Swer
28.	*Helicia robusta* Wall.	Proteaceae	+	+
29.	*Ilex embeloides* Hook.f.	Aquifoliaceae	+	+
30.	*Ilex theaefolia* Hk.f.	Aquifoliaceae	–	+
31.	*Ilex venulosa* Hk.f.	Aquifoliaceae	+	+
32.	*Itea chinensis* Hook and Arn.	Iteaceae	+	+
33.	*Itea macrophylla* Wall.	Iteaceae	–	+
34.	*Ligustrum lucidum* Aiton.	Oleaceae	–	+
35.	*Ligustrum robustum* Bl.	Oleaceae	–	+
36.	*Lindera latifolia* Hk.f.	Lauraceae	+	+
37.	*Lithocarpus dealbatas* Hk.f. & Th. Mss.	Fagaceae	+	–
38.	*Litsea salicifolia* Roxb.	Lauraceae	+	+
39.	*Lyonia ovolifolia* (Wall.) Drude.	Ericaceae	–	+
40.	*Magnolia pterocarpa* Roxb.	Magnoliaceae	+	+
41.	*Maesa indica* Wall.	Myrsinaceae	–	+
42.	*Mussaenda roxburghii* Hk.f	Rubiaceae	+	–
43.	*Myrica esculenta* Buch.-Ham. ex D.Don	Myricaceae	+	+
44.	*Myrsine semiserrata* Wall.	Myrsinaceae	+	+
45.	*Neolitsea cassia* (Linn.) Koster.	Lauraceae	+	+
46.	*Pentapanax subcordatum* (G.Don) Hutch.	Araliaceae	+	+
47.	*Persea duthei* (Hook.f.) Koster.	Lauraceae	+	+
48.	*Persea odoratissima* (Nees) Koster.	Lauraceae	–	+
49.	*Phoebe lanceolata* Nees.	Lauraceae	–	+
50.	*Photinia integrifolia* Lindl.	Rosaceae	+	+
51.	*Pinus kesiya* Royle ex Gordon	Pinaceae	–	+
52.	*Prunus acuminata* Wall.	Rosaceae	–	+
53.	*Prunus jenkensii* Hk.f.& Th.	Rosaceae	+	+
54.	*Quercus glauca* Thunb.	Fagaceae	+	–
55.	*Rhododendron arboreum* Sm.	Ericaceae	+	+
56.	*Schefflera elata* (Buch.-Ham.) Harms	Araliaceae	+	+
57.	*Schefflera hypoleuca* Kurz. Harms	Araliaceae	+	+

Contd...

Table 5–Contd...

Sl.No.	Name of Species	Family	Sohrarim	Swer
58.	*Schima khasiana* Dyer.	Theaceae	+	–
59.	*Symplocos glomerata* King ex Gamble	Symplocaceae	+	+
60.	*Symplocos* sp.	Symplocaceae	–	+
61.	*Symplocos pyrifolia* Wall. Ex G. Don.	Symplocaceae	+	+
62.	*Symplocos spicata* Roxb.	Symplocaceae	+	+
63.	*Syzygium tetragonum* (Wt.) Kurz.	Myrtaceae	+	+
64.	*Viburnum simonsii* Hk.f.	Caprifoliaceae	+	+
II.	**Lianas**			
1.	*Derris marginata* Benth.	Fabaceae	–	+
2.	*Hedera helix* Cl.	Araliaceae	–	+
3.	*Holboellia latifolia* Wall.	Lardizabalaceae	–	+
4.	*Melodinus monogynus* Roxb.	Apocynaceae	+	+
5.	*Piper* sp.	Piperaceae	–	+
6.	*Tetrastigma serrulatum* (Roxb.) Planch	Vitaceae	+	+
7.	*Tupidanthus calyptratus* Hk.f. and Th.	Araliaceae	–	+
III.	**Shrubs**			
1.	*Ardisia oxyphylla* Wall.	Myrsinaceae	+	+
2.	*Berberis willichiana* D.C.	Berberidaceae	+	–
3.	*Coffea jenkinsii* Hk.f.	Rubiaceae	+	–
4.	*Coffea khasiana* Hk.f.	Rubiaceae	+	–
5.	*Elaeagnus latifolia* Linn.	Eleagnaceae	–	+
6.	*Euonymus theifolius* Wall.ex Laws.	Celastraceae	+	+
7.	*Goniothalamus sesquipedalis* Hk.f. and Th.	Annonaceae	+	–
8.	*Hedyotis scandens* Roxb.	Rubiaceae	+	–
9.	*Ixora nilgricans* Bl.	Rubiaceae	–	+
10.	*Jasminum calycinum* Wall.	Oleaceae	+	–
11.	*Lasianthus lancifolius* Hk.f.	Rubiaceae	+	+
12.	*Lasianthus lucidus* Blume.	Rubiaceae	+	+
13.	*Macrosolen psilanthus* (Hk.f.) Danser	Loranthaceae	+	–

Contd...

Table 5–Contd...

Sl.No.	Name of Species	Family	Sohrarim	Swer
14.	*Piper khasianum* Cas. DC.	Piperaceae	+	+
15.	*Psychotria symplocifolia* Kurz.	Rubiaceae	+	+
16.	*Sarcococca pruniformis* Lindl.	Buxaceae	+	+
17.	*Senecio cappa* Buch.-Ham.ex D.Don	Asteraceae	+	+
18.	*Smilax* sp.	Smilacaceae	+	+
19.	*Strychonos wallichiana* Benth.	Loganiaceae	+	–
20.	*Vaccinium griffithianium* Wt.	Vacciniaceae	+	+
21.	*Vaccinium serratum* Wight.	Vaccinaceae	+	–
22.	*Viburnum foetidum* Wall.	Caprifoliaceae	+	–
IV.	**Herbs**			
1.	*Alyxia fascicularis* Benth.	Apocynaceae	+	–
2.	*Anoetochilus roxburghii* (Wall.) Lindl.	Orchidaceae	+	+
3.	*Arisaema tortuosum* (Wall.) Scott.	Araceae	+	+
4.	*Balanophora dioca* Brown.	Balanophoraceae	+	+
5.	*Begonia lacinata* Roxb.	Begoniaceae	+	–
6.	*Botrychium laguninosum* Wall. Ex Hook et Grev.	Polypodiaceae	+	–
7.	*Carex baccans* Nees.	Cyperaceae	+	–
8.	*Dicranopteris* sp.	Pteridaceae	+	+
9.	*Globba* sp.	Zingiberaceae	+	+
10.	*Gleichenia* sp.	Gleicheniaceae	+	+
11.	*Hedychium spicatum* Lodd.	Zingiberaceae	+	–
12.	*Hemiphragma heterophyllum* Wall.	Scrophulariaceae	–	+
13.	*Impatiens* sp.	Balsaminaceae	+	–
14.	*Jasminium* sp.	Oleaceae	+	+
15.	*Monotropa uniflora* Linn.	Monotropaceae	+	–
16.	*Ophiopogon parviflorus* (Hook.f.) Hara	Liliaceae	+	+
17.	*Paspalum orbiculare* Frost.	Poaceae	+	–
18.	*Pteris* sp.	Pteridaceae	+	–
19.	*Rhaphidophora lancifolia* Schott.	Araceae	+	–
20.	*Rubia cordifolia* Linn.	Rubiaceae	–	+

Contd...

Table 5–Contd...

Sl.No.	Name of Species	Family	Sohrarim	Swer
21.	*Rubus niveus* Thunb. Diss.	Rosaceae	+	–
22.	*Rubus rugosus* Sm.	Rosaceae	+	–
23.	*Sarcandra glabra* (Thunb.) Nakai.	Chloranthaceae	+	–
24.	*Sellaginella* sp.	Sellaginellaceae	+	–
25.	*Viola tricolor* Linn.	Violaceae	+	+

+: Presence; –: Absence.

Table 6: Wild Edible Fruits Bearing Species Found in the Sacred Groves of the Jaintia Hills of Meghalaya (After Upadhaya *et al.*, 2005).

Plant Species	*Local Name*	*Sacred Grove*
Caryota urens Linn.	Kwai-cha	1 and 2
Castanopsis indica A.DC.	Deing-sarang	1 and 2
Castanopsis purpurella (Mq) Balak	Deing–soh-stap	1 and 2
Castanopsis tribuloides (Sm.) DC.	Deing Soh-ot	1 and 2
Cayratia japonica (Thunb.) Gagnep	Syrbuid	1 and 2
Citrus latipes (Swingle) Tanaka	Soh-heh	1 and 2
Diospyros kaki Thunb.	Dieng-iong	1 and 2
Docynia indica (Wall) Decne	Sa-phoh-btet	1
Elaeagnus pyriformis Hk.f.	Soh-shang	1 and 2
Garcinia tinctoria (DC.) W.F. Wight	Soh-khyllung	1 and 2
Holboellia latifolia Wall.	Sa-tymbra	2
Meyna spinosa Link	Sa-matan	1 and 2
Myrica esculanta Buch-Ham.	Soh-phie	1 and 2
Passiflora edulies Sims.	Soh-brab	1 and 2
Prunus jenkensii Hk.f.	Sa-tanghi	1 and 2
Pseudostreblus indica Bureau	Deing -chirikhlaw	1
Pyrularia edulis A.DC.	Deing Soh-Klong	1 and 2
Pyrus pashia D.Don	Soh-shur	1
Rhus javanica Linn.	Sa-ma	1 and 2
Spondias axallaris Roxb.	Sa-liat	1 and 2
Syzygium tetragonum (Wt.) Kurz	Deing-soh-sarlei	1 and 2
Tetrastigma leucostaphylum (Dennst) Balak.	Syr-pung	2
Tetrastigma serrulatum (Roxb.) Planch.	Sla-ngar	1
Viburnum foetidum Wall.	Syrlei	1 and 2
Viburnum simonsii Hk.f. and Th.	Syrlei-bri	1
Zanthoxylum acanthopodium DC.	Jaiur	1

1: Ialong sacred grove; 2: Raliang sacred grove.

Table 7: Wild Leafy Vegetables Used by the Jaintia Tribe of Meghalaya (After Upadhaya *et al.*, 2005).

Plant Species	*Local Name*	*Sacred Grove*
Arisaema consanguineum Schott	Saru-bsein	1 and 2
Begonia palmata D.Don	Sla-lajaw	1 and 2
Centella asiatica (L.) Urban	Tyngkhieh	1 and 2
Corylopsis himalayana Griff.	Laphaiur	1 and 2
Cyathea gigantea (Wall. ex Hk.) Holttum	Deing tyrkhang	1 and 2
Dysoxylon gobara (Buch.-Ham) Merr.	Sla-luchai	2
Ficus elmerii Merr.	Deing-thylling-sang	1
Houttuynia cordata Thunb.	Jamyrdoh	1 and 2
Plantago erosa Wall.	Shkur-blang	1 and 2
Polygonum dibotrys D.Don.	Jarain	1 and 2
Rhododendron arboreum Sm.	La-ku-thiin	1
Saurauia punduana Wall.	Sla-mehek	1 and 2
Schima wallichii (DC.) Korth.	Shyr-ngan	1 and 2
Tetrastigma leucostaphylum (Dennst) Balak.	Slangnar	2
Tetrastigma serrulatum (Roxb.) Planch.	Syr-pung	1
Zanthoxylum acanthopodium DC.	Jaiur	1

1: Ialong sacred grove; 2: Raliang sacred grove.

Table 8: Medicinal Plants Found in Ialong and Raliang Sacred Groves of Jaintia Hills of Meghalaya (After Upadhaya *et al.*, 2005)

Plant Species	*Local Name*	*Parts Used*	*Uses*	*Sacred Grove*
Achyranthes aspera L.	Sabyrthit	Leaf/stem	Extract is used for teeth problem	1 and 2
Aeschynanthes superba Cl.	Thylli-masi	Bark	Paste is used in bone fracture and muscular sprain (external use only)	2
Agapetes auriculata Hk.f.	Tynchang khian	Stem/bark	Crushed, made into a paste and applied on fractured bone/dislocation of joints (external use only)	1 and 2
Agapetes variegata (Roxb) G.Don	Tynchang hek	Swollen stem/ bark	The paste made of stem and bark is applied in bone fracture/dislocation of joints (external use only)	1 and 2
Alstonia scholaris (Linn.) R	Deing-ryteng	Bark	Extract is used for cold and cough, fever, dysentery and gastrointestinal problems	2
Arisaema consanguineum Schott	Saru-bsein	Root	Paste is used for skin diseases (external use only)	1 and 2
Artemisia nilagirica (Cl.) Pamp.	Byrlud	Leaf	Leaf paste is used as antifungal agent in cuts and wounds	1 and 2
Asparagus filicinus	Chana-rieh sla Buch-Ham	Root	Extract is used in indigestion and as appetizer	2
Begonia palmata D.Don.	Sla-lajaw	Root/stem/leaf	Eaten raw in indigestion, fever and extract is used for eye injury	1 and 2

Contd...

Table 8–Contd...

Plant Species	*Local Name*	*Parts Used*	*Uses*	*Sacred Grove*
Cassia mimosoides L	–	Whole plant	Paste is applied in skin disease, cuts and wounds	1 and 2
Centella asiatica (L.) Urb.	Tyngkh-ieh	Whole plant	Eaten raw in dysentery, diarrhea and high blood sugar	1 and 2
Cirsium involucratum DC.	Ch-eih-keng	Root	Used in gastro-intestinal disorder, and as appetizer	1 and 2
Citrus latipes (Swingle) Tanaka	Soh-heh	Fruit/young leaf	Used in body ache, reduce vomiting, weakness, cold, fever and acts as appetizer	1 and 2
Corylopsis himalayana Griff	La-pha-iur	Flower/leaf buds	Used in diarrhea, dysentery and gastro-intestinal disorder	1 and 2
Costos specious (Koenig.) Smith	Krah-hei-iang	Whole plant	Extract is used in skin diseases, leprosy, asthma, bronchitis, anemia and inflammations	1 and 2
Cyathea gigantea (Wall. ex Hk.) Holttum.	Dein-tyrkhang	Young leaf	Used in weakness and mental stress.	1 and 2
Disporum cantoniense (Lour.) Merr.	–	Root	Crushed and the extract is used for stomach problem	1 and 2
Docynia indica (Wall) Decaisne	Sa-phoh-btet	Fruit/young leaf	Fruits are used against cold; cough, indigestion and leaves are used in dysentery and diarrhea	1

Contd...

Table 8–Contd...

Plant Species	*Local Name*	*Parts Used*	*Uses*	*Sacred Grove*
Drymaria cordata (L) Roem and Schult	Phlang-bsein	Whole plant	Used as antidote for snakebite, and mental disorders	1 and 2
Dysoxylum gobara (Buch-Ham) Merr.	Sla-luchai	Young leaf	Boiled and used in malaria, fever and blood pressure	2
Eriosema himalaicum Ohashi	Sa-pyrdong	Tuberous root	Used in intestinal worms, indigestion and also acts as appetizer	1
Hedychium sp.	Sy-ein-kloo	Rhizome/root	Eaten raw or crushed with water and used in gastric ulcer, asthma and skin diseases	1 and 2
Hedyotis scandens D.Don	Tyllie-iumurem	Whole plant	The extract is used in cold and cough, rheumatism, kidney problem, muscles sprain and fracture	1 and 2
Helixanthera ligustrina (Wall) Danser	Mang-ka-riang	Bark of swollen stem	Crushed and used in gastric problem	1
Hodgsonia macrocarpa Bl	Sma-khang	Whole plant	Boiled or crushed and used in gastric ulcer and malaria	1 and 2
Houttuynia cordata Thunb	Ja-myrdoh	Whole plant	Eaten raw as salad in iron deficiency and blood sugar	1 and 2
Hoya parviflora Wight.	-	Bark of swollen stem	Paste is applied in bone fracture/dislocation of joints (external use only)	1 and 2

Contd...

Table 8–Contd...

Plant Species	*Local Name*	*Parts Used*	*Uses*	*Sacred Grove*
Lycopodium cernum Linn	Tmain-khla	Root/stem	Used as first aid in cuts and wounds to stop bleeding	1 and 2
Mahonia pycnophylla (Fedde) Takeda	Deing–la-ranong	Leaf/bark	Crushed and the juice is diluted with water and used in various eye disease	1
Meyna spinosa Link	Sa-matan	Fruit	External application for healing of cuts and cracks	1 and 2
Myrica esculanta Buch-Ham.	Sa-phai	Bark	Crushed, boiled and used in ringworm	1 and 2
Osbeckia capitata Walp.	Sa-pythem	Whole plant	Used as antidote in snakebite and swelling of muscles	1
Oxalis corniculata L	Phlang-kba	Whole plant	Used in fever, headache, mental stress and eye problems	1 and 2
Paederia scandens (Lour.) Merr.	Phlang-khinang	Whole plant	Used in indigestion, dysentery and diarrhea	1 and 2
Piper griffithii C.DC	Pathi-bri	Leaf/fruit/stem	Crushed and used in boils and skin disease	1 and 2
Plantago erosa Wall	Shkurblang	Leaf	Used as ointment in cuts and burns.	1 and 2
Polygonum dibotrys D.Don	la-rain	Leaf/stem	Rubbed against itching caused by caterpillar and insect sting, skin disease	1 and 2
Pouzolzia hirta (Bl) Hassk.	Tyngknor	Root	Used in tumor, swelling of muscles and skin diseases	1 and 2

Contd...

Table 8–Contd...

Plant Species	*Local Name*	*Parts Used*	*Uses*	*Sacred Grove*
Raphidophora sp.	Dawai-rarbek	Stem/leaf	Crushed, made into a paste and wrapped around the fractured part and swollen muscles	1 and 2
Rhus javanica Linn.	Sa-ma	Fruit	Crushed fruit is used in stomach and gastric problems	1 and 2
Sida rhombifolia (L)	Sokha-khnei	Leaf	Juice mixed with mustured oil for massaging.	1 and 2
Stephania hernandifolia (Willd) Walp	Longdong-sla	Whole plant	Extract used in kidney problem and as appetizer	1 and 2
Tetrastigma leucostaphylum (Dennst) Balakr.	Syrpung	Leaf/bark	Extract is believed to be a good tonic for liver and give relieves from fever and body pain	2
Toddalia asiatica (L) Lamk.	Phlang-chieh	Leaf/root/bark	Extract used in kidney and stomach problems	1 and 2
Zanthoxylum oxyphyllum Edgew	Jaiur-blai	Leaf/fruit	Extract is rubbed to cure skin disease and toothache	1 and 2

1: Ialong sacred grove; 2: Raliang sacred grove.

Table 9: Endemic (E), Rare (R) and Threatened (T) Plant Species Found in Sacred Groves of Jaintia Hills of Meghalaya

Sl.No	*Plant Species*	*Family*	*Status*	*Distribution*
1.	*Acer laevigatum* Wall.	Aceraceae	E, R	Temperate Himalaya, Sikkim and Meghalaya
2.	*Acer oblongum* Wall.	Aceraceae	R	Indo-Malaya, Himalayas and North East India
3.	**Aeschynanthes parasiticus* (Roxb.) Wall.	Gesneriaceae	E	Endemic to Meghalaya
4.	*Aeschynanthes sikkimensis* (Clarke) Stapf.	Gesneriaceae	E, R	Sikkim, North East India
5.	**Aeschynanthes superba* Clarke.	Gesneriaceae	E	Endemic to Meghalaya
6.	*Aralia thomsonii* Seem	Araliaceae	E	Eastern Himalayas
7.	*Ardisia disperma* Cl.	Myrsinaceae	R	Eastern Himalaya to Burma
8.	**Ardisia griffithii* Cl.	Myrsinaceae	E	Endemic to Meghalaya
9.	*Ardisia odontophylla* DC.	Myrsinaceae	R	Burma and North East India
10.	*Balanophora dioca* Royle	Balanophoraceae	R	Sub-Himalayas, North-East India, Indo-Burma and Nepal
11.	**Baliospermum micranthum* Muell-Arg.	Euphorbiaceae	E	Endemic to Meghalaya
12.	*Boehmeria sidaefolia* Wedd.	Urticaceae	E	Indo-Malaya
13.	*Bruceae mollis* Wall. ex Kurz.	Semaroubiaceae	R	South-East Asia, North- East India and Andamans

Contd...

Table 9–Contd...

Sl.No	Plant Species	Family	Status	Distribution
14.	*Bulbophyllum griffithii* (Lindl.) Reiclb	Orchidaceae	E, R	Sikkim and North-East India
15.	*Callicarpa psilocalyx* Clarke	Verbenaceae	E, R	North-East India
16.	**Camellia caduca* Cl. ex Brandis	Theaceae	E	Endemic to Meghalaya
17.	*Carpinus viminea* Lindl.	Corylaceae	E	Temperate Himalaya and Burma
18.	*Capparis acutifolia* Sweet	Capparaceae	E	Endemic to North-East India
19.	*Ceropegia angustifolia* Wt.	Asclepiadaeceae	E, R	North-East India
20.	*Cinnamomum pauciflorum* Nees	Lauraceae	E, R	North-East India
21.	**Citrus latipes* (Swingle) Tanaka	Rutaceae	E, R	Endemic to Meghalaya
22.	*Croton oblongus* Burm.f.	Euphorbiaceae	R	Indo-Malaya
23.	*Cyathea* gigantea (Wall ex Hook.) Holttm	Cyatheaceae	R, T	North-East India
24.	*Dendrobium densiflorum* Wall.	Orchidaceae	R	Sikkim, North-East India, Indo-Nepal
25.	*Dendrobium devonianum* Paxt.	Orchidaceae	R	North-East India, Sikkim, Bhutan, Burma and Thailand
26.	*Dendrobium nobile* Lindl.	Orchidaceae	R	North-East India, Sikkim, Bhutan, Nepal, Thailand and China
27.	*Dipteris wallichii* (R.Br.) Moore	Dipteridaceae	R	North East India
28.	*Drimycarpus racemosus* (Roxb.) Hook.f.	Anacardiaceae	E	Eastern Himalayas and Bangladesh

Contd...

Table 9–Contd...

Sl.No	Plant Species	Family	Status	Distribution
29.	*Drosera peltata* Smith	Droseraceae	R	Himalayas, Indo-Malaya, Nilgiris and Australia
30.	*Embelia vestita* Roxb.	Myrsinaceae	R	North-East India
31.	*Erythroxylum kunthianum* Wall. ex Kurz	Erythroxylaceae	E	Indo-Burma, North-East India
32.	**Euonymus lawsonii* Clarke and Prain	Celastraceae	E	Endemic to Meghalaya
33.	*Ficus concinna* Miq.	Moraceae	R	North-East India
34.	*Ficus subincisa* Buch.-Ham. Ex J.E.SH	Moraceae	R	Himalaya to Burma
35.	*Fissistigma verrucosum* (Hk.f.&Th) Merr.	Annonaceae	E, R	Endemic to North-East India
36.	*Fraxinus floribunda* Wall.	Oleaceae	R	Temperate and Sub-Himalaya
37.	*Glochidion thomsonii* (Muell-Arg.) Hook. f.	Euphorbiaceae	E	Bangladesh, North-East India
38.	*Gnetum montanum* Mark Graf	Gnetaceae	R	Eastern Himalayas, North-East India, South India, Indo-China
39.	**Gomphostemna lucidum* Benth.	Lamiaceae	E, R	Endemic to Meghalaya
40.	**Goldfussia glabrata* (Nees) Balakr.	Acanthaceae	E	Endemic to Meghalaya
41.	*Hedera nepalensis* K.Koch	Araliaceae	E, R	Bhutan Himalayas
42.	**Ilex embeloides* Hook.f.	Aquifoliaceae	E, R	Endemic to Meghalaya
43.	**Ilex venulosa* Hook.f.	Aquifoliaceae	E, R	Endemic to Meghalaya
44.	**Impatiens acuminata* Hook.f.	Balsaminaceae	E	Endemic to Meghalaya

Contd...

Table 9–Contd...

Sl.No	Plant Species	Family	Status	Distribution
45.	**Impatiens juripa* Hook. f. and Th	Balsaminaceae	E	North-East India
46.	**Impatiens khasiana* Hook.f. and Th	Balsaminaceae	E	Endemic to Meghalaya
47.	**Impatiens laevigatum* Hook.f. and Th	Balsaminaceae	E	Endemic to Meghalaya
48.	**Impatiens porrecta* Hook. f. and Th	Balsaminaceae	E	Endemic to Meghalaya
49.	*Ixora subsessiles* G.Don	Rubiaceae	E	North-East India
50.	*Leucosceptrum canum* Smith	Lamiaceae	R	Temperate and Sub-Himalayas
51.	**Lindera latifolia* Hook.f.	Lauraceae	E, R	Endemic to Meghalaya
52.	*Litsea leata* (Nees) Hook.f.	Lauraceae	E	Bangaladesh and Eastern Himalayas
53.	*Luisia inconspicua* (Hook.f) King and Pantl.	Orchidaceae	E, R	Sikkim, North-East India
54.	*Mahonia pycnophylla* (Fedde) Takeda	Berberidaceae	E	Indo-Burma, Eastern Himalayas and Nilgiris
55.	*Manglietia insignis* Bl.	Magnoliaceae	R	Eastern Himalayas
56.	*Melodinus monogynous* Roxb.	Apocynaceae	R	Indo-Malaya, North-East India
57.	*Morinda umbellate* Linn.	Rubiaceae	R	Burma, Bangladesh, Himalaya and Sub-Himalaya
58.	*Munronia pinnata* (Wall.) Harms.	Meliaceae	R	Tropical Himalaya and Nilgiris
59.	*Neillia thyrsiflora* D.Don	Rosaceae	E	Indo-Malaya and Himalayas
60.	*Osbekia capitata* Benth.	Melastomaceae	E	East Bhutan, Meghalaya

Contd...

Table 9–Contd...

Sl.No	*Plant Species*	*Family*	*Status*	*Distribution*
61.	**Paramignya micrantha* Kurz	Rutaceae	E	Endemic to Meghalaya
62.	**Persea parviflora* (Meissn).Haridasan et Rao	Lauraceae	E	Endemic to Meghalaya
63.	*Phlogancanthus pubinervius* T. Andes	Acanthaceae	E	Eastern Himalayas
64.	*Piper griffithii* C.DC.	Piperaceae	E	North-East India
65.	*Piper peepuloides* Roxb.	Piperaceae	E	Himalayas, North-East India
66.	*Pleione maculata* (Lindl.) Lindl.	Orchidaceae	R	North-East India, Sikkim, Bhutan, Nepal and Thailand
67.	*Pleione praecox* (Lindl.) Lindl.	Orchidaceae	R	North-East India, Sikkim, Bhutan, Nepal, Burma and Thailand
68.	*Podocarpus neriifolia* D.Don	Podocarpaceae	R	North-East India, Indo-Burma, Malaya and Japan
69.	*Pseudobrassiopsis hispida* (Seem.) R.N.Ban	Araliaceae	R	Burma and North-East India
70.	*Pyrularia edulis* A. DC.	Santalaceae	R	Temperate and Subtropical Himalayas
71.	**Pogostemon strigosus* (Benth.) Benth.	Lamiaceae	E	Endemic to Meghalaya
72.	*Porana racemosa* Roxb.	Convolvulaceae	E, R	Subtropical Himalayas to Burma
73.	*Prunus jenkinsii* Hook.f.	Rosaceae	E	North-East India
74.	*Psychotria symplicifolia* Kurz.	Rubiaceae	E, R	Burma, North East India
75.	*Quercus glauca* Thunb.	Fagaceae	R	Subtropical Himalayas and Japan

Contd...

Table 9–Contd...

Sl.No	Plant Species	Family	Status	Distribution
76.	*Rapidophora calophyllum* Schott.	Araceae	E	Endemic to North East India
77.	*Rapidophora decursiva* (Roxb.) Schott.	Araceae	E	Sikkim and North-East India
78.	*Rhus hookerii* Sahni and Bahd.	Anacardiaceae	R	Sikkim Himalayas and Meghalaya
79.	*Rubus assemensis* Focke	Rosaceae	E	Burma, North-East India
80.	**Rubus khasianus* Cordat	Rosaceae	E	Endemic to Meghalaya
81.	*Sarcosperma griffithii* Cl.	Sapotaceae	R	North-East India
82.	**Schima khasiana* Dyer.	Theaceae	E, R	Endemic to Meghalaya
83.	**Senecio jowaiensis* Balakr.	Asteraceae	E	Endemic to Meghalaya
84.	*Smilax myrtillus* DC.	Smilacaceae	E	North-East India
85.	**Sonerila khasiana* Cl.	Melostomaceae	E, R	Endemic to Meghalaya
86.	*Styrax hookerii* Cl.	Styracaceae	R	Eastern and Sub-Himalayas
87.	**Sympagia monodelpha* (Nees) Bremek.	Acanthaceae	E	Endemic to Meghalaya
88.	**Tupidanthus calyptratus* Hk.f. and Th.	Araliaceae	E, R	Endemic to Meghalaya
89.	*Turpina nepalensis* W. and A. Prodr.	Staphylaceae	E, R	Indo-Malaya
90.	*Vaccinium vacciniaceum* (Roxb.) Sleum.	Vacciniaceae	E	North-East India
91.	**Viburnum simonsii* Hk.f. and Th.	Caprifoliaceae	E	Endemic to Meghalaya

Sources: Chauhan (1983), Kataki (1983), Kumar and Rao (1985), Haridasan and Rao (1983), Takhtajan (1988), Khan *et al.* (1997), Upadhaya (2002), Jamir and Pandey (2003), Pandey *et al.* (2005).

*: Endemic to Meghalaya.

Table 10: Rare, Endangered and Threatened Category of Plants Found in Sacred Grove in Khasi Hills of Meghalaya (R: Rare; E: Endemic; EN: Endangered; CR: Critically endangered; V: Vulnerable). Species are rare to Meghalaya and endemic either to the state of Meghalaya or northeast India or Eastern Himalayas (After Rao and Haridasan 1983, Kamei 2007, Nongbri 2007, Barik *et al.*, 2008).

Species	*Family*	*Status*	*Sacred Groves*
Acer laevigatum Wall.	Aceraceae	R	Shillong peak, Mawphlang
Aeschynanthus sikkimensis (Cl.) Stapf.	Gesneriaceae	R	Shllong peak
Agapetes obovata (Wt.) Hk.f.	Vacciniaceae	E	Mawsmai
Agapetes rugosus Hk.f.	Vacciniaceae	E	Shillong peak
Anthoxanthum clarkei Hk.f.	Poaceae	EN	Nongkrem
Apios cornea Benth.	Fabaceae	R	Shillong peak
Balliospermum micranthus Muell-Arg.	Euphorbiaceae	R, E	Mawsmai, Mawphlang
Calanthe manii Hook.f.	Orchidaceae	R	Mawphlang, Shillong peak
Camellia caduca Cl.ex Brandis	Theaceae	E	Mawsynram
Clematis apiculata Hook.f.& Th.	Ranunculaceae	R	Cherrapunji
Coffea jenkinsii Hk.f.	Rubiaceae	E	Mawsmai
Dendrobium devonianum Paxt.	Orchidaceae	R	Cherrapunji
Diplomeris pulchella Don.	Orchidaceae	R, E	Cherrapunji, Sohrarim
Elaeocarpus acuminatus Wall. ex Mast	Elaeocarpaceae	R	Mawsmai
Elaeocarpus prunifolius C. Muell. Mast	Elaeocarpaceae	R	Mawsmai

Contd...

Table 10–Contd...

Species	Family	Status	Sacred Groves
Epipogum rosceum (D.Don) Lindl.	Orchidaceae	R	Mawphlang
Eria ferruginea Lindl.	Orchidaceae	E	Cherrapunji
Eria pusilla (Griff.) Lindl.	Orchidaceae	E	Cherrapunji
Euonymus bullatus Wall. ex Lindl.	Celastraceae	R	Mawsmai
Goodyera recurua Lindl.	Orchidaceae	E	Mawphlang
Habenaria khasiana Hook.f	Orchidaceae	E	Cherrapunji
Hedera helix L.	Araliaceae	R	Shillong peak, Mawphlang
Hoya acuminata Benth.	Asclepiadaceae	R	Mawsmai
Ilex khasiana Purk.	Aquifoliaceae	R, E,CR	Shillong peak
Ilex venulosa Hook.F.	Aquifoliaceae	R, E	Sohrarim
Lasianthus biermannii King ex Hk.F.	Rubiaceae	R	Sohrarim
Leptodermis griffithii Hk. f.	Rubiaceae	R	Mawphlang
Ligustrum compactum Hk.f and Th.	Oleaceae	R	Mawsmai
Ligustrum confusum Hk.f and Th.	Oleaceae	R	Sohrarim
Lindera latifolia Hk.f.	Lauraceae	R, E	Shillong peak
Litsea elongata Wall.	Lauraceae	R	Nongkrem
Luvunga scandens Ham.	Rutaceae	R	Mawsmai

Contd...

Table 10–Contd...

Species	*Family*	*Status*	*Sacred Groves*
Manglietia caveana Hk.f.& Th.	Magnoliaceae	R	Mawphlang
Medinella rubicunda Bl.	Melastomaceae	R	Mawsmai
Mitrastemon yamamotoi (Makino)	Mitrastemonaceae	EN	Mawsmai
Monotropa uniflora L.	Monotropaceae	R	Mawphlang
Neanotis oxyphylla (G. Don) W.H.Lewis	Rubiaceae	E	Mawsmai
Panax pseudo-ginseng Wall.	Araliaceae	V	Shillong peak, Mawphlang
Pauthrea arguta Decne	Rosaceae	R	Shillong peak
Pentapterigium rugosum Hk.f.	Vacciniaceae	R, E	Mawphlang, Shillong peak
Persea kingii (Hk.f.) Koster	Lauraceae	E	Shillong peak
Photinia integrifolia Lindl.	Rosaceae	R	Cherrapunji, Swer
Polygala tricholopha Chodat	Polygalaceae	E	Mawsmai
Porana racemosa Roxb.	Convolvulaceae	R	Shillong peak
Psychotria symplocifolia Kurz.	Rubiaceae	R	Sohrarim
Schima khasiana Dyer.	Theaceae	R	Shillong peak, Sohrarim
Schizandra neglecta Sm.	Schizandraceae	R	Shillong peak
Smilacina fusca Wall.	Liliaceae	R	Mawphlang
Sophora acuminata Benth. ex. Baker	Fabaceae	R	Mawsmai

Contd...

Table 10–Contd...

Species	*Family*	*Status*	*Sacred Groves*
Sophora wightii Baker	Fabaceae	R	Mawsmai
Thelasis bifolia Hook.f.	Orchidaceae	R	Cherrapunji
Viburnum simonsii Hk.f.& Th.	Caprifoliaceae	E	Mawsynram, Swer

Bibliography

Abrams MD (1990). Adaptations and responses to drought in *Quercus* species of North America. *Tree Physiology* 7: 227-238.

Aiba, Shin-ichiro and Kitayama K (1999). Structure, composition and species diversity in an altitude-substrate matrix of rain forest tree communities on Mount Kinabalu, Borneo. *Plant Ecology* 140: 139-157.

Anonymous (1974). *Geology and Mineral Resources of the States of India. GSI*, Miscellaneous Publication 30: 124.

Anonymous (1984). *Meghalaya Forests*. Chief conservator of forest, Meghalaya, Shillong, India.

Anonymous (1996). *Scared Groves of Andhra Pradesh*. World Wide Fund for Nature, Andhra Pradesh State Office, Hyderabad.

Anonymous (1999). *State of Forest Report*. Forest Survey of India, Dehradun.

Armesto JJ, Mutchell JD and Villogram C (1986). A comparison of spatial patterns of trees in tropical and temperate forests. *Biotropica* 18: 1-11.

Armstrong DP and Westoby M (1993). Seedlings from large seeds tolerate defoliation better: a test using phytogenetically independent contrasts. *Ecology* 74:1092-1100.

Ashton PS and Hall P (1992). Comparison of structure among mixed dipterocarp forest of North-Western Borneo. *Journal of Ecology* 80: 459-481.

Ashton PS (1969). Speciation among tropical forest trees: some deductions in the light of recent evidence. *Biology Journal of Linnaean Society London* 1: 155–96.

Ayyappan N and Parthasarathy N (1999). Biodiversity inventory of trees in a large-scale permanent plot of tropical evergreen forest at Varagalaiar, Anamalais, Western Ghats, India. *Biodiversity and Conservation* 8: 1533-1554.

Baker HG (1972). Seed weight in relation to environmental conditions in California. *Ecology* 53:007-1010.

Balakrishnan NP (1981-1983). *Flora of Jowai.* 2 Vols. Botanical Survey of India, Howrah.

Balasubramanyam, K and Induchoodan, NC (1996). Plant diversity in sacred groves of Kerala. *Evergreen* 36:3-4.

Barik SK (1992). *Ecology of Tree Regeneration Along a Disturbance Gradient in a Subtropical Wet Hill Forest of Meghalaya.* Ph.D. Thesis, North-Eastern Hill University, Shillong, Meghalaya.

Barik SK, Haridasan K and Lakadong NJ (2008). Medicinal plant resources of Meghalaya: Endemism, threat status and consumption pattern. *Envis Forestry Bulletin* 7 (2): 17-25.

Barik SK, Pandey HN, Tiwari, BK and Singh B (2006). *Sacred groves of Meghalaya: A scientific and conservation perspective.* Regional Centre, National Afforestation and Eco-Development Board, North-Eastern Hill University, Shillong.

Barik SK, Pandey HN, Tripathi RS and Rao P (1992). Microenvironmental variability and species diversity in tree fall gaps in a subtropical broadleaved forest. *Vegetatio* 103: 31-40.

Barik, S.K., Tripathi RS, Pandey, HN and Rao P (1996a). Tree regeneration in a subtropical humid forest: effect of cultural disturbance on seed production and germination. *Journal of Applied Ecology* 33:1551-1560.

Barik, SK, Rao P, Tripathi RS, and Pandey HN (1996b). Dynamics of tree sedling populations in a humid subtropical forest of northeast India as related to disturbance. *Canadian Journal of Forest Research* 26:584-589.

Barnes BV, Zak DR, Denton SR and Spurr H (1998). *Forest Ecology*, John Wiley and Sons, Inc. New York.

Basu R (2000). Studies on sacred groves and taboos in Purulia district of West Bengal. *Indian Forester* 1309-1317.

Bauhas J, Pare D and Cote L (1998). Effects of tree species and stand age and soil type on soil microbial biomass and it activity in a southern boreal forest. *Soil Biology and Biochemistry* 30: 1077-1089.

Benzing DH (1983). Vascular epiphytes: a survey with special reference to their interactions with other organisms. In: Sutton SL, Whitmore TC and Chadwick AC (eds.) *Tropical Rain Forest: Ecology and Management*. Blackwell Scientific Publication, Oxford.

Bhakat RK (1990). Tribal ethics of forest conservation. *Yojana* (March 16-31): 23-27.

Bhuyan P, Khan ML and Uma Shankar (2000). Trade-off between dispersal efficiency and seedling fitness in *Oroxylum indicum*, a wind dispersed tropical tree. *International Journal of Ecology and Environmental Sciences* 26: 67-73.

Bonfil C (1998). The effect of seed size, cotyledon reserves and herbivory on seedling survival and growth in *Quercus rugosa* and *Q. laurina* (Fagaceae). *American Journal of Botany* 85: 79-87.

Boojh R and Ramakrishnan PS (1993). Studies on sacred groves and their role in environmental conservation. In: *Strategies for Environmental Management*, Souvenir volume, Department of Science and Environment of Uttar Pradesh, Lucknow, pp. 6-8.

Box EO (1995). Climatic relations of the forest of East and South-East Asia. In: Box EO, Peet RK, Masuzawa T, Yamada I, Fujiwara K, and Maycock PF (eds.) *Vegetation Science in Forestry*. Kluwer Academic Publishers, Dordretch, Netherlands, pp. 23-55.

Brandis D (1897a). Indian Forestry, Working Oriental Institute, UK.

Brandis D (1897b). Indigenous Indian Forestry: Sacred Groves in Indian Forestry. Working Oriental Institute, UK. pp. 12-13.

Brokaw NVL (1985). Tree falls, regrowth and community structure in tropical forests. In: Pickett STA and White PS (eds.) *The Ecology of Natural Disturbance and Patch Dynamics*. Academic Press, New York, pp.53-69.

Brown JH (1981). Two decades of homage to Santa Roselia: towards a general theory of diversity. *American Zoologist* 21: 877-888.

Burman RJJ (1992). The institution of sacred grove. *Journal of Indian Anthropological Society* 27: 219-238.

Campbell DG, Stone JL and Rosas A Jr. (1992). A comparison of the phytosociology and dynamics of three flood plain (Varzae) forests of known ages, Rio Jurua, Western Brazilian Amazon. *Botanical Journal of Linnaean Society* 108:213-237.

Cao Kun-Fang and Rob P (1997). Species diversity of Chinese beech forests in relation to warmth and climate disturbances. *Ecological Research* 12: 175-189.

Champion HG (1936). A preliminary survey of forest types of India and Burma. *Indian Forest Record* 1: 1-286.

Champion HG and Seth SK (1968). *A Revised Survey of the Forest Type of India*. Government of India, New Delhi.

Chauhan AS and Singh DK (1992).Changing pattern in the flora of Meghalaya due to deforestation. In Gupta A and Dhar DC (eds.) *Environment Conservation and Wasteland Development in Meghalaya*. Meghalaya Science Society, Shillong.

Chauhan AS (1983). Dwindling taxa of Meghalaya. In: Jain SK and Rao RR (eds.) *An Assessment of Threatened Plants of India*. Botanical Survey of India, Howrah, pp. 142-145.

Clark DB and Clark DA (1987). Population ecology and microhabitat distribution of *Dipteryx panamensis*, a neotropical rainforest tree. *Biotropica* 19: 236-244.

Collins SL, Glenn SM and Gibson DJ (1995). Experimental analysis of intermediate disturbance and initial floristic composition: decoupling cause and effect. *Ecology* 76: 486-492.

Connell JH (1978). Diversity in tropical rain forest and coral reefs. *Science* 199: 1302-1310.

Connell JH and Lowman MD (1989). Low-diversity tropical rain forests: Some possible mechanisms for their existence. *The American Naturalist* 134: 89–119.

Dabadghao PM and Shankarnarayan KA (1973). *The Grass Cover of India* ICAR, New Delhi.

Daniels RJR, Gadgil M and Joshi MV (1995). Impact of human extraction on tropical humid forests in Western Ghats in Uttar Kannada, South India. *Journal of Applied Ecology* 32:866-874.

Das AK, Boral L, Tripathi RS and Pandey HN (1997). Nitrogen mineralization, microbial biomass-N in a subtropical humid forest of Meghalaya, India. *Soil Biology and Biochemistry* 29: 1609-1612.

Davis TAW and Richards PW (1933-34). The vegetation of Moraballi Creek, British Guiana: an ecological study of a limited area of tropical rain forest. I and II. *Journal of Ecology* 21: 350-84, 22:106-55.

Denslow JS (1980). Gap partitioning among tropical rain forest trees. *Biotropica* 12 (Supplement): 47-59.

Devi S (2000). Sacred groves of Manipur. Abstract, National Workshop on Community Strategies on the Management of Natural Resources, Bhopal, India.

Ehrlich, JH and Wilson, EO (1991). Biodiversity studies: science and policy. *Science* 253:758-762.

Felfili JM (1995). Diversity, structure and dynamics of a gallery forest in central Brazil. *Vegetatio* 117: 1-15.

Fenner M (1991). Irregular seed crops in forest trees. *Quarterly Journal of Forestry* 85:166-172.

Finzi AC, Van Breeman N, and Canham CD, (1998). Canopy tree-soil interactions within temperate forests: tree species effects on carbon and nitrogen. *Ecological Applications* 8: 440-446.

Foster SA (1986). On the adaptive value of large seeds for tropical moist forest trees: a review and synthesis. *Botanical Review* 52:261-299.

Foster SA and Janson SA (1985). The relationship between seed size and establishment conditions in tropical woody plants. *Ecology* 66:773-780.

Gadgil M and Vartak VD (1975). Sacred groves of India: A plea for continued conservation. *Journal of Bombay Natural History Society* 72:314-320.

Gadgil R and Vartak VD (1976). The sacred groves of Western Ghats in India. *Economic Botany* 30: 152-160.

Gadgil M and Vartak VD (1981). Sacred groves of Maharashtra: An inventory. In: Jain SK (ed.) *Glimpses of Ethnobotany*. Oxford University Press, Bombay, pp. 279-294.

Ganesh T, Ganesan R, Soubadra Devy M, Davidar P and Bawa KS (1996). Assessment of plant biodiversity at a mid-elevation evergreen forest of Kalakad-Mundanthurai Tiger Reserve, Western Ghats, India. *Current Science* 71(5): 379-392.

Ganeshaiah KN and Uma Shaanker R (1991). Seed size optimization in a wind dispersed tree *Butea monosperma*: A trade-off between seedling establishment and pod dispersal efficiency. *Oikos* 60: 3-6.

Ganashaiah KN, Shaanker RV and Bawa KS (1997). Diversity of species assemblages of Islands: Predictions and their test using tree species composition of Shola fragments. *Current Science* 73(2): 188-194.

Gentry AH (1988). Tree species richness of upper Amazon forests. *Proceedings of National Academy of Sciences, USA* 85: 156-159.

Gerden C and Mtallo S (1990). Traditional Forest Reserves in Babati District, Tanzania: A study in human ecology. Working Paper 128, Swedish University of Agricultural Sciences, IRDC, Uppsala.

Greene DF and Jhonson EA (2004). Modeling the temporal variation in seed production of north American trees. *Canadian Journal of Forest Research* 34: 65-75.

Gross KL (1984). Effect of seed size and growth form on seedling establishment of six monocarpic perennial plants. *Journal of Ecology* 72: 369-387.

Gross KL and Kromer ML (1986). Seed weight effects on growth and reproduction in *Oenothera bennis* L. *Bulletin Torrey Botanical Club* 113:252-258.

Grubb PJ (1977). Control of forest growth and distribution of wet tropical mountains with special reference to mineral nutrition. *Annual Review of Ecology and Systematics* 8:83-107.

Grubb PJ and Burslem DFRP (1998). Mineral nutrient concentrations as a function of seed size within seed crop: implications for competition among seedlings and defense against herbivory. *Journal of Tropical Ecology* 14: 177-185.

Grubb PJ and Coomes DA (1997). Seed mass and nutrient content in nutrient-starved tropical rainforest in Venezuela. *Seed Science Research* 7: 269:280.

Grubb PJ (1996). Rain forest dynamics: the need for new paradigms. In: Edwards DS, Booth WE and Choy SC (eds) *Tropical Rainforest Research: Current Issues.* Kluwer, Dordrecht, pp. 215-233.

Hajra PK (1975). Law Lyngdoh or Sacred groves at Mawphlang, Visitor's guide, Government of Meghalaya, Shillong.

Hajra PK (1980). Nature Conservation in Khasi folk beliefs and taboos. In: Jain SK (ed.) *Glimpses of Indian Ethnobotany.* Oxford and IBH, New Delhi. pp. 149-152.

Hamilton MB (1999). Tropical tree gene flow and dispersal. *Nature* 401: 129.

Haridasan K and Rao RR (1983). Threatened plants of Meghalaya– aplea for conservation. In: Jain SK and Rao RR (eds.) *An Assessment of Threatened Plants of India.* Botanical Survey of India, Howrah, pp. 94-103.

Haridasan K and Rao RR (1985-1987). *Forest Flora of Meghalaya.* 2 Vols. Bishen Singh Mahendrapal Singh, Dehradun, India.

Harms KE, Wright SJ, Calderon O, Hernandez A and Herre A (2000). Pervasive density-dependent recruitment enhances seedling diversity in a tropical forest. *Nature* 404: 493-495.

Harper JL (1977). *The Population Biology of Plants.* Academic press, New York, USA.

Hedge SG, Ganeshaiah KN and Uma Shaanker R (1991). Fruit preference criteria by avian frugivores: their implication for the evolution of clutch-size in *Solanum pubescence. Oikos* 6:20-26.

Hooper D and Vitousek PM (1997). The effects of plant composition and diversity on ecosystem processes. *Science* 277: 1302-1305.

Horn H (1966). The measurement of overlap in comparative ecological studies. *American Naturalist* 100:419-424.

Howe HF (1982). Fruit production and animal activity at two tropical trees, In: Leigh Jr. E, Rand A S and Windsor D (eds.) *The Ecology of Tropical Rain Forest.* Smithsonian Institution Press, Washington DC, pp. 189-200.

Hubbell SP and Foster RB (1983). Canopy gaps and the dynamics of neotropical forest. In: Crawley MJ (ed.) *Plant Ecology*, Blackwell Sci. Publication, Oxford, pp. 77-96.

Hubbell, SP (1979). Tree dispersion, abundance and diversity in a tropical dry forest. *Science* 203: 1299-1309.

Huges JD (1984). Sacred groves: The Gods, forest protection and sustained yield in the ancient world. In: Steen HK (ed.) *History of Sustained yield Forestry: A Symposium*. Forest History Society, Durham, North Carolina, pp. 331-343.

Hughes JD and Chandran MDS (1998). Sacred groves around the Earth: An Overview. In: Ramakrishnan PS, Saxena KG and Chandrashekaran UM (eds.) *Conserving the Sacred Groves for Biodiversity Management*, Oxford and IBH Publishing Co. Pvt. Ltd., New Delhi.

Induchoodan, NC (1996). *Ecological studies on the sacred groves of Kerala*. PhD Thesis, Central University of Pondicherry, India.

Jaccard P (1912). The distribution of flora of Alpine zone. *New Phytologist* 11: 37-50.

Jamir SA and Pandey HN (2003). Vascular plant diversity in the sacred groves of Jaintia Hills in northeast India. *Biodiversity and conservation* 12:1497-1510.

Jamir SA (2000). *Studies on Plant Biodiversity, Community Structure, and Population Behaviour of Dominant Tree Species in some Sacred Groves of Jaintia hills, Meghalaya*. PhD Thesis. North-Eastern Hill University, Shillong.

Jamir SA, Upadhaya K and Pandey H N (2006). Life-form composition and stratification of montane humid forests in Meghalaya, northeast India. *Tropical Ecology* 47(2) 183-190.

Janzen DH (1971). Seed predation by animals. *Annual Review of Ecology and Systematics* 2: 465-492.

Jayarajan J (2004). *Sacred Groves of Malabar*. Kerala Research Programme on Local Development, Centre for Developmental Studies, Prashant Nagar, Ulloor, Thiruvananthapuram, India.

Jensen TS (1985). Seed-seed predator interactions of European beech, *Fagus sylvatica* and forest rodents, *Clethrionomys glareolus* and *Apodemus flavicollis*. *Oikos* 44:149-156.

John B, Pandey HN and Tripathi RS (2001). Vertical distribution and seasonal changes of fine and coarse root mass in *Pinus kesiya* Royle ex. Gordon forest of three different ages. *Acta Oecologica* 22:293-300.

Johnston M and Gillman M (1995). Tree population studies in low-diversity forest, Guyana. I. Floristic composition and stand structure. *Biodiversity and Conservation* 4: 339-463.

Kadamba D (1998). *Biocultural Perspective and Plant Diversity of Sacred Groves and Traditional Medicinal Knowledge in Pondicherry*. PhD Thesis, Pondicherry University, India.

Kadavul K and Parthasarathy N (1999). Structure and composition of woody species in tropical semi-evergreen forest of Karajan hills, Eastern Ghats, India. *Tropical Ecology*, 40: 247-260.

Kalam MA (1996). Sacred groves in Kodagn district of Karnataka, Pondy. Paper on social sciences, French Institute, Pondicherry.

Kamei J (2007). *Studies on Interrelationship between Tree Diversity and N and P Dynamics in a Humid Subtropical Forest Ecosystem of Meghalaya*. Ph.D. Thesis, North-Eastern hill University, Shillong.

Kanjilal UN, Kanjilal, PC, Das A, De RN and Bor N L (1934–40). Flora of Assam, Vols. 1–5, Government Press, Shillong.

Kataki SK (1983). Some rare plants in Khasi and Jaintia hills of Meghalaya. In: Jain SK and Rao RR (eds.) *An Assessment of Threatened Plants of India*, Botanical Survey of India, Howrah, pp. 146-150.

Khan ML (1986). *Ecological Studies on Regeneration of Few Tree Species in Sub-tropical Wet Hill Forests of Meghalaya*. PhD Thesis. North-Eastern Hill University, Shillong.

Khan ML, Bhuyan P, Uma Shankar, Sing ND and Todaria NP (1999). Seed germination and seedling fitness in *Mesua ferra* L. in relation to fruit size and seed number per fruit. *Acta Oecologica* 20:599-606.

Khan ML, Khumbongmayum AD and Tripathi RS (2008).The sacred groves and their significance in conserving biodiversity: An overview. *International Journal of Ecological and Environmental Sciences* 34: 277-291.

Khan ML, Rai JPN and Tripathi RS (1986). Regeneration and survival of tree seedlings and sprouts in tropical deciduous and sub-tropical forest of Meghalaya, India. *Forest Ecology and Management* 14: 293-304.

Khan ML, Shaily M, and Bawa KS (1997). Effectiveness of the protected area network in biodiversity conservation, a case study of Meghalaya state. *Biodiversity and Conservation* 6: 853–868.

Khan ML and Uma Shankar (2001). Effect of seed weight, light regime and substratum microsite on germination and seedling growth of *Quercus semiserrata* Roxb. *Tropical Ecology* 42:117-125.

Khiewtam RS and Ramakrishnan PS (1993). Litter and fine root dynamics of relic sacred grove forest of Cherrapunji in northeastern India. *Forest Ecology and Management* 60: 327-344.

Khiewtam RS (1986). *Ecosystem Function of Protected Forests of Cherrapunji and Adjoining Area.* Ph.D. Thesis, North-Eastern Hill University, Shillong.

Khumbongmayum AD, Khan ML and Tripathi RS (2005). Survival and growth of seedlings of a few tree species in the four sacred groves of Manipur, Northeast India. *Current Science* 88 (11): 1781-1785.

Kikizawa K (1988). Dispersal of *Quercus mongolica* acorns in a broad leaved deciduous forest 1. Disappearance. *Forest Ecology and Management* 25:1-8

Kumar Y and Rao RR (1985). Studies on Balphakram Wildlife Sanctuary in Meghalaya, 3. General account, forest types and fauna. *Indian Journal of Forest Research* 8: 300-309.

Kushalapa CG, Bhagwat SA and Kushalapa KA (2001). Conservation and management of sacred groves of Hodagu, Karnataka, South India- a unique approach. In: KN Ganeshaiah, R Uma Shaanker and KS Biswas (eds.) *Tropical Ecosystems: Structure, Diversity and Human Welfare*, Oxford IBH Publishing Co. Pvt. Ltd., New Delhi, pp. 565-569.

Lal JB (1989). *India's Forests: Myth and Reality*. Natraj Publishers, New Delhi, India.

Law P (2002). *Studies on Population Ecology of Keystone Species and Their Role in Ecosystem Functions in the Two Sacred Groves of Meghalaya.* Ph.D. Thesis, North-Eastern Hill University, Shillong.

Lawton JH and Brown VK (1994). Redundancy in ecosystems. In: Schulze ED and Mooney HA (eds.) *Biodiversity and Ecosystem Function.* Springer Verlag, Berlin, Heidelberg, pp. 255-270.

Lieberman D, Lieberman M, Peralta R and Hartshorn GS (1996). Tropical forest structure and composition on a large-scale altitudinal gradient in Costa Rica. *Journal of Ecology* 84: 137-152.

Loreau (2000) Biodiversity and ecosystem functioning: recent theoretical advances. *Oikos* 91: 3-17.

Lovett G and Reuth H (1999). Soil nitrogen transformations in beech and maple stands along a nitrogen deposition gradient. *Ecological Applications* 9:1330-1344.

Magurran A (1988). *Ecological Diversity and its Measurement.* Princeton University Press, New Jersey.

Majumdar SK (1986). The pre-Cambrian framework of part of the Khasi hills, Meghalaya. GSI 117, pp. 59.

Malhotra K C (1998). Anthropological dimensions of sacred groves in India: An overview. In: Ramakrishnan PS, Saxena KG and Chandrashekaran UM (eds.) *Conserving the Sacred Groves for Biodiversity Management,* Oxford and IBH publishing Co. Pvt. Ltd. New Delhi, pp. 423-438.

Malhotra KC, Gogkhle Y, Chatterjee S and Srivastava S (2001). *Cultural and Ecological Dimensions of Sacred Groves in India.* Indian National Science Academy and Indira Gandhi Rashtriya Manav Sangrahalaya, Bhopal.

Malhotra KC, Gogkhle Y, Das K (2001a). *Sacred groves of India: Annotated Bibliography.* Indian National Science Academy and Development Alliance, New Delhi.

Malhotra KC, Gogkhle Y, Chatterjee S and Srivastava S (2007). *Sacred Groves in India an Overview,* Aryan Books International, New Delhi.

Malhotra KC, Stanley S, Hemam NS and Das K (1997). Biodiversity conservation and ethics: Sacred groves and pools. In: Fujiki N and Macer J (eds.) *Bioethics in Asia*. Eubios Institute, Japan pp. 338-45.

Manokaran N, La frankie JV, Kochumman KM, Quah ES, Klahn J, Ashton PS and Hubbell, SP (1991). Stand table and distribution of species in the 50 ha reserve plot at Pasoh forest reserve. Forest Research Institute of Malaysia, Reserve data.

Marshell DL (1986). Effect of seed size on seedling success in three species of *Sesbania* (Fabaceae). *American Journal of Botany* 73:457-464.

Martinez-Ramos M and Soto-Castro A (1993). Seed rain and advanced regeneration in a tropical rain forest. *Vegetatio* 107: 299-318.

Melillo JM, Aber JD and Muratore JF (1982). Nitrogen and lignin control of hardwood leaf litter decomposition dynamics. *Ecology* 63: 621-626.

Myers N, Mittermeir RA, Mittermeir CJ, da Fonseca GAB and Kents J (2000). Biodiversity hotspots for conservation priorities. Nature 403:853-858.

Mishra BP, Tripathi OP, Tripathi RS and Pandey HN (2004). Effect of anthropogenic disturbance on plant diversity and community structure of a sacred grove in Meghalaya, northeast India. *Biodiversity and Conservation* 13: 421-436.

Monselise SP and Goldschmidt EE (1982). Alternate bearing in fruit trees. *Horticultural Review* 4: 128-173.

Morisita M (1959). Measuring of interspecific association and similarity between communities. *Mem. Fac. Sci.Kyushu Univ. Ser. E* (Biol) 3: 65-80.

Murli KS (1997). Pattern of seed size, germination and seed viability of tropical tree species in Southern India. *Biotropica* 29: 271-279.

Myers N, Mittermiller CG, Gustava AB, Da Foseca and Kent J (2000). Biodiversity hotspots for conservation priorities. *Nature* 403:853-858.

Nadkarni NM, Matelson TJ and Haber WA (1995). Structural characteristics and floristic composition of neotropical cloud forest, Monteverde, Costa Rica. *Journal of Tropical Ecology* 11: 482-495.

Nongbri, S (2007). *Studies on Regeneration of Tree Species in Subtropical Forests of Meghalaya.* Ph.D. Thesis, North-Eastern Hill University, Shillong.

Norse EA and McManus RE (1980). Ecology and living resources biological diversity. In: *Environmental Quality 1980: The Eleventh Annual Report of Council on Environmental Quality.* Council of Environmental Quality, Washington, D.C, pp. 31-80.

Odyuo N (2004). *Ecological Studies of Some Important Tree Species of Subtropical Humid Forest of Meghalaya.* Ph.D. Thesis, North Eastern Hill University, Shillong.

Okuda T, Naoki K, Yap SK and Manokaran N (1997). Tree distribution pattern and fate of juveniles in a lowland tropical rain forest–implication for regeneration and maintenance of species diversity. *Plant Ecology* 131: 155-171.

Panda D, Kumar PP and Das AP (2003). Ten important sacred groves of Santhals in the Bankura district of West Bengal, India. Abstract, XIII Annual Conference of Indian Association for Angiosperm Taxonomy and International Symposium on Plant Taxonomy: Advances and Relevance, Deptt. of Botany, TM Bhagalpur University, Bhagalpur, India.

Pandey HN, Tripathi OP and Tripathi RS (2003). Ecological analysis of forest vegetation of Meghalaya. In: Bhatt BP, Bujarbaruah KM, Sharma YP and Patiram (Eds.) *Approaches for Increasing Agricultural Productivity in Hill and Mountain Ecosystem* ICAR Research Complex for N.E.H. Region, Shillong pp.37-50.

Pandey HN, Upadhaya K, Jamir SA, Law PS and Tripathi RS (2005). Floristic diversity in the sacred groves of Meghalaya. In: Pandey A.K, Jun Wen and Dogra JVV (Eds.) *Plant Taxonomy: Advances and Relevance*, CBS Publishers and Distributors, New Delhi pp. 83-99.

Parrish, JAD and Bazzaz FA (1982). Niche responses of early and late successional tree seedlings on three resource gradients. *Bulletin Torrey Botanical Club* 109:451-456.

Parthasarathy N and Karthikeyan R (1997). Plant biodiversity inventory and conservation of two tropical dry evergreen forests on the Coromandal Coast, South India. *Biodiversity and Conservation* 6: 1063-1083.

Parthasarathy N (1987). Seasonal dynamics of fine roots in a tropical forest in South India. *Journal of Indian Botanical Society* 66:338-345.

Pascal JP (1992). Evergreen forests of the Western Ghats: Structural and functional trends. In: Singh KP and Singh JS (eds) *Tropical Ecosystems: Ecology and Management.* Wiley Eastern Limited, New Delhi, pp. 385–408.

Perez-Gracia F, Iriondo JM, Martinez-Laborde JB (1995). Germination behavior in seeds of *Diplotaxis erucoides* and *D. virgata. Weed Research* 35: 495-502.

Persson H (1983). The distribution and productivity of fine roots in boreal forest. *Plant and Soil* 71:87-101.

Pimm SL (1984). The complexity and stability of ecosystems. *Nature* 307: 312-326.

Poore MED (1968). Studies in Malaysian forest, I. The forest in Triassic sediments in Jengka forest reserve. *Journal of Ecology* 56:143-196.

Puri GS (1960). *Indian Forest Ecology,* Vol. I and II. Oxford Book and Stationery. Co., New Delhi.

Puri GS, Meher-Homji VM, Gupta RK and Puri GS (1983). *Forest Ecology*. Vol. I and II. Oxford and IBH Publishing Co. Pvt. Ltd., New Delhi.

Putman RJ (1994). Community Ecology. Chapman and Hall, London.

Rao AS (1974). The vegetation and phytogeography of Assam-Burma. In: Mani MS (ed.) *Ecology and Biogeography in India.* W. Junk publishers, The Hague, pp. 204–205.

Rao P (1992). *Ecology of Gap Phase Regeneration in a Subtropical Broad-leaved Climax Forest of Meghalaya.* Ph. D. Thesis, North-Eastern Hill University, Shillong, Meghalaya.

Rao P (1996). Sacred grove and conservation. WWF-India Quarterly 7: 4-8.

Rao P, Barik SK, Pandey HN and Tripathi RS (1997). Tree seedling germination, establishment in tree fall gaps and understorey in a subtropical forest of northeast India. *Australian Journal of Ecology* 22: 136-145.

Rao RR and Hajra PK (1986). Floristic diversity of Eastern Himalaya in a conservation perspective. *Proceedings of Indian Academic of Sciences* (Animal Science/Plant Science) Suppl. 103-125.

Rao RR and Haridasan K (1983). Threatened plants of Meghalaya– A plea for conservation. In: Jain SK and Rao RR (eds.) *An Assessment of Threatened plants of India*, Botanical Survey of India, Howrah pp. 94-103.

Rao RR (1994). *Biodiversity in India: Floristic aspects.* Bishen Singh Mahendra Pal Singh, Dehradun, India.

Raunkiaer C (1934). *The Life Forms of Plants and Statistical Plant Geography*. Clarendon Press, Oxford, England.

Richards PW (1996). *Tropical Rain Forest, an Ecological Study.* Cambridge University Press, Cambridge. U.K.

Roy S and Singh JS (1994). Consequences of habitat heterogeneity for availability of nutrients in dry tropical forest. *Journal of Ecology* 82:503-509

Runkle JR (1981). Gap regeneration in some old-growth forests of eastern United States. *Ecology* 62: 1041-1051.

Sahunulu P and Dhanmononda P (1995). Structure and dynamics of dry dipterocarp forest, Sakaerat, North-Eastern Thailand. In: Box EO, Peet RK, Masuzawa T, Yamada I, Fujiwara K, and Maycock PF (eds.) *Vegetation Science in Forestry*. Kluwer Academic Publishers, Dordretch, Netherlands, pp.465-495.

Schulze ED and Mooney HA (1993). *Biodiversity and Ecosystem Function*. Springer-Verlag, Berlin, Germany.

Schupp EW (1990). Annual variation in seed fall, post-dispersal, dispersal predation and recruitment of a Neotropical tree. *Ecology* 71: 504-515.

Shannon CE and Weiner W (1963). *The Mathematical Theory of Communication*. University of Illinois press, Urbana, pp.177.

Simpson EH (1949). Measurement of diversity. Nature 163: 688.

Singh S (1996). *Technical report on delineation of geo-ecological zones in Meghalaya.* Report No. GE/UGC/SR (1)/SS, Department of Geography, North-Eastern Hill University, Shillong.

Slobodkin LB and Sanders HL (1969). On the contribution of environmental predictability to species diversity. *Brookhaven Symposia in Biology* 22:82-95.

Song Youngchang (1995). The global position of the evergreen broad-leaved forest of China. In: Box EO, Peet RK, Masuzawa T, Yamada I, Fujiwara K, and Maycock PF (eds.) *Vegetation Science in Forestry*. Kluwer Academic Publishers, Dordretch, Netherlands, pp. 69-84.

Sorensen J (1948). A method of establishing groups of equal amplitude in plant sociology based on similarity of species content. Detkong. Dauske Vidrnsk. *Selsk. Biol. Skn.* 5 (4): 1-34.

Srivastava SK, Singh KP and Upadhaya (1986). Fine root growth dynamics in teak (*Tectona grandis* Linn. F.). *Canadian Journal Forest Research* 16:1360-1364.

Sukumar R, Dattaraja HS, Suresh HS, Radakrishnan JV, Vasudeva R, Nirmala S and Joshi NV (1992). Long-term monitoring of vegetation in a tropical deciduous forest in Mudumalai, southern India. *Current Science* 62: 608-616.

Sundrapandian SM and Swamy PS (1996). Influence of herbaceous species composition on fine root biomass production in disturbed deciduous forests of Western Ghats in India. *Acta Oecologia* 17:163-176.

Sundrapandian SM, Chandrasekharan S and Swamy PS (1999). Variation in fine root mass and net productivity due to conversion of tropical forests into forest plantations. *Tropical Ecology* 40:305-312.

Takhtajan A (1969). *Flowering plants, Origin and dispersal.* Edinburg.

Takhtajan A (1988). *Floristic Regions of the World*. Bishen Singh Mahandra Pal Singh, Dehradun, pp. 71-78.

Thompson J, Proctor V, Vianna V, Milliken W, Ratter JA and Scott DA (1992). Ecological studies on a low land evergreen rain forest at Maraca Island, Roraima, Brazil. I. Physical environment, forest structure and leaf chemistry. Journal of Ecology 80: 689-703.

Tiedje JM (1988). Ecology of denitrification and dissimilarity nitrate reduction to ammonium In: Zehnder AJB (ed.) *Biology of Anaerobic Microorganisms*, John Wiley, New York.

Tilman D, Knops J, Wedin D, Reich P, Ritchie M. and Siemann E (1997). The influence of functional diversity and composition on ecosystem processes. *Science* 227: 1300-1302.

Tiwari BK, Barik SK and Tripathi RS (1999). *Sacred Forests of Meghalaya. Biological and Cultural Diversity.* Regional Centre, National Afforestation and Eco-Development Board, North Eastern Hill University, Shillong.

Tripathi RS and Khan ML (1990). Effects of seed weight and microsite characteristics on germination and seedling fitness in two species of *Quercus* in a subtropical wet hill forest. *Oikos* 57: 289-296.

Turner IM and Corlett RT (1996). The conservation value of small, isolated fragments of lowland tropical rain forest. *Trends in Ecology and Evolution*, 11:330-333.

Upadhaya K (2002). *Studies on Plant Biodiversity and Ecosystem Function in Sacred Groves of Meghalaya.* Ph. D. Thesis, North-Eastern Hill University, Shillong.

Upadhaya K, Pandey HN, Law PS and Tripathi RS (2003). Tree diversity in sacred groves of Jaintia Hills in Meghalaya, northeast India. *Biodiversity and Conservation* 12: 583-597.

Upadhaya K, Pandey HN, Law PS and Tripathi RS (2004). Diversity and population characteristics of woody species in subtropical humid forests exposed to cultural disturbances in Meghalaya, northeast, India. *Tropical Ecology* 45: 303-314.

Upadhaya K, Pandey HN, Law PS and Tripathi RS (2005). Dynamics of fine and coarse roots and nitrogen mineralization in a humid subtropical forest ecosystem of northeast India. *Biology and Fertility of Soils* 41: 144-152.

Upadhaya K, Pandey HN, Law PS and Tripathi RS (2005a). Plants of ethenobotanical importance in the sacred groves of Jaintia Hills of Meghalaya. *The Indian Forester* 131(6): 819-828.

Upadhaya K, Pandey HN and Tripathi RS (2006). Understory plant diversity in subtropical humid forest of Meghalaya. *International Journal of Ecology and Environmental Sciences* 32:207-219.

Upadhaya, K, Barik SK, Pandey HN and Tripathi OP (2008).

Response of woody species to anthropogenic disturbances in sacred forests of northeast India. *International Journal of Ecology and Environmental Sciences* 34:245-257.

Upadhaya, K, Pandey HN and Law PS (2007). The effect of seed mass on germination, seedling survival and growth in *Prunus jenkinsii* Hook.f. and Thoms. *Turkish Journal of Botany* 31:31-36.

Vazquezg J, Thomas A and Givinish J (1998). Altitudinal gradients in tropical forests composition, structure and diversity in the Sierra de Manantan. *Journal of Ecology* 86: 999-1020.

Visalakshi N (1995). Vegetation analysis of two tropical dry evergreen forests of South India. *Tropical Ecology* 36: 117-127.

Vitousek PM, Gosz JR, Grier CC, Melillo JM and Reiners W (1982). A comparative analysis of potential nitrification and nitrate mobility in forest ecosystem. *Ecological Monographs* 52:155-177.

Weih M and Karlsson PS (1999). The nitrogen economy of mountain birch seedlings: implications for winter survival. *Journal of Ecology* 87:211–219.

Welden CW, Hewett SW, Hubbell SP and Foster RB (1991) Sapling survival growth and recruitment: relationship to canopy height in a neotropical forest. Ecology 72: 35-50.

Whitmore TC (1975). *Tropical Rain Forests of the Far East*. Oxford University Press, Oxford.

Whitmore TC (1996). A review of some aspects of tropical rain forest seedling ecology with suggestions for further enquiry. In: Swaine MD (ed.) *The Ecology of Tropical Forests Tree Seedlings*. Parthenon Publishing Group, Paris, France.

Whittaker RH (1972). Evolution and measurement of species diversity. *Taxon* 21:213-251.

Whittaker RH (1975). Communities and Ecosystems. Mac Millan Publishing Co, New York.

Winn AA and Werner PA (1987). Regulation of seed yield within and among populations of *Prunella vulgaris*. *Ecology* 68:1224-1233.

Zak DR and Pregitzer KS (1990). Spatial and temporal variability of nitrogen cycling in northern Lower Michigan. *Forest Science* 36:367-380.

Index

L

M

N

O

P